Insecticides

Detection, Management, and Environmental Impact

nova science publishers

www.novapublishers.com

Environmental Research Advances Series

Biodiversity in Forest Ecosystems: Patterns, Drivers, and Conservation Strategies
Niamjit Das
ISBN: 979-8-89530-882-0
eBook ISBN: 979-8-89530-917-9

Public Health: Understanding the Impact of Environmental Pollutants
Emmanuel Obeng-Gyasi, PhD, MPH and Joav Merrick, MD, MMedSci, DMSc
ISBN: 979-8-89530-579-9
eBook ISBN: 979-8-89530-636-9

Water, Pesticides, Organisms: Current Approach of Ecotoxicology
Marilia Hartmann, PhD, Paulo Afonso Hartmann, PhD, Caroline Müller, PhD and Aline Pompermaier, PhD
ISBN: 979-8-89530-284-2
eBook ISBN: 979-8-89530-390-0

Tipping the Boundaries: Health and Well-Being of Sustainable Development
Sandeep Poddar, PhD and Waliza Ansar, PhD
ISBN: 979-8-89530-146-3
eBook ISBN: 979-8-89530-378-8

Illegal, Unreported, and Unregulated (IUU) Fishing: Impacts, Costs, and Countermeasures
Lorenzo Scherpiani and Stefano Rea
ISBN: 979-8-89530-084-8
eBook ISBN: 979-8-89530-191-3

Socio-Environmental Change: The Path Toward Sustainability
Adele Sateriano, PhD, Alessia D'Agata and Ioannis Vardopoulos
ISBN: 979-8-89530-124-1
eBook ISBN: 979-8-89530-139-5

More information about this series can be found at
https://novapublishers.com/product-category/series/environmental-research-advances/

Archana Anokhe
Govindaraj Kamalam Dinesh
Editors

Insecticides

Detection, Management, and Environmental Impact

nova
science publishers
www.novapublishers.com

DOI: https://doi.org/10.52305/ZSFQ1341

Library of Congress Cataloging-in-Publication Data

ISBN: 979-8-90134-022-6 (Hardcover)
ISBN: 979-8-90134-114-8 (eBook)

Published by Nova Science Publishers, Inc. † New York

Contents

Abbreviations

AChE	acetylcholine esterase
AHS	Agricultural Health Study
AICRP	All India Coordinated Research Project
AINP-PR	All-India Network Project on Pesticide Residues
ALP	alkaline phosphatase
ALS	amyotrophic lateral sclerosis
AQI	air quality index
AOPs	advanced oxidation processes
ADHD	attention deficit hyperactivity disorder
BChE	butyrylcholinesterase
BR	Bioremediation
CAGR	compound annual growth rate
CBF	carbofuran
CCD	colony collapse disorder
CCPR	Codex Committee on Pesticide Residues
CIBRC	Central Insecticides Board and Registration Committee
cVOCs	chlorinated volatile organic compounds
CE	Capillary Electrophoresis
DAD	diode array detector
DA&FW	Department of Agriculture and Farmers' Welfare
DART-MS	direct analysis in real-time mass spectrometry
DESI-MS	desorption electrospray ionization mass spectrometry
DIA	dipstick immunoassay
DISPME-GC-MS	direct immersion-solid-phase microextraction – gas chromatography-mass spectrometry
ECD	electron-capture detector
EKT	Electrokinetic treatments

ELISA	enzyme-linked immunosorbent assay
EPA	Environmental Protection Agency
EKT	Electrokinetic treatments
FO	forward osmosis
FRET	Fluorescence Resonance Energy Transfer
FSSAI	Food Safety and Standards Authority of India
FPD	flame photometric detector
GABA	gamma-aminobutyric acid
GAP	Good Agricultural Practices
GC	gas chromatography
GHS	Globally Harmonized System
GMOs	genetically modified organisms
GNP	gold nanoparticle
GPC	Gel permeation chromatography
HCH	hexachlorocyclohexane
HMs	Heavy metals
HPLC	high-performance liquid chromatography
HPG	high-pressure gradient
HRP	horseradish peroxidase
ICA	immunochromatographic assay
ICAR	Indian Council of Agricultural Research
IEC	Ion exchange chromatography
IPM	integrated pest management
IoT	Internet of Things
IT	ion trap
JMPR	Joint Meeting on Pesticide Residues
KOH	potassium hydroxide
LC-MS/MS	Liquid Chromatography–Tandem Mass Spectrometry
LLE	Liquid-Liquid Extraction
LOD	limit of detection
LOQ	limit of quantification
LPG	low-pressure gradient
LTP-MS	low-temperature plasma mass spectrometry
MAE	Microwave-assisted extraction
MAPE	magnetic solid-phase extraction
MCL	Maximum Containment Level

MEKC	Micellar electrokinetic capillary chromatography
MOF	Metal-organic framework
MPRNL	Monitoring of Pesticide Residues at National Level
MRLs	maximum residue limits
MS	Mass Spectrometry
MS (other)	Mass Spectroscopy
MSD	mass selective detector
MSPD	Matrix-solid-phase dispersion
MS/MS	tandem mass spectrometry
nAChR	nicotinic acetylcholine receptor
NBR	Nano-bioremediation
NPD	nitrogen phosphorus detector
NF	Nanofiltration
NHL	non-Hodgkin lymphoma
NPs	Nanoparticles
nZVI	Nano zero-valent iron
OHS	occupational health and safety
OPPs	organophosphorus pesticides
PAHs	polycyclic aromatic hydrocarbons
PCB	polychlorinated biphenyl
PCBs	polychlorinated biphenyls
PDA	photodiode array
PFF	profenofos
PIC	Prior Informed Consent
PIPs	plant-incorporated protectants
PLE	Pressurised liquid extraction
POPs	persistent organic pollutants
PS-MS	paper spray mass spectrometry
PTs	Pesticides
Q-TOF	quadrupole time-of-flight mass spectrometer
Q-Trap	quadrupole linear ion trap mass analyzer
QQQ / QqQ	triple quadrupole analyzer
QuEChERS	Quick, Easy, Cheap, Effective, Rugged and Safe method
RPLC	Reversed phase chromatography
RO	reverse osmosis
SEC	size exclusion chromatography

SFC	Supercritical Fluid Chromatography
SiNPs	siliconized silica nanoparticles
SPE	Solid-phase extraction (if needed)
SPME	Solid-phase micro-extraction
SVE	Soil vapour extraction
TEPP	Tetraethyl Pyrophosphate
TLC	thin-layer chromatography
TOF / TOF-MS	time-of-flight (mass spectrometry)
TrOCs	trace organic compounds
UHPLC	ultra-high-performance liquid chromatography
USEPA	United States Environmental Protection Agency
UV-Vis	UV-Visible spectroscopy
WHO	World Health Organisation

Chapter 1

Overview of Chemical Insecticides: History, Usage, and Environmental Impacts

Samrat Saha[1*]
Archana Anokhe[2,*]
and Govindaraj Kamalam Dinesh[3*]
[1]Department of Agricultural Entomology, Uttar Banga Krishi Viswavidyalaya, Pundibari, Cooch Behar, West Bengal, India
[2]ICAR - Directorate of Weed Research, Jabalpur, Madhya Pradesh, India
[3] Department of Biochemistry, Physiology, Microbiology and Environmental Science, College of Agriculture, Central Agricultural University, Imphal, India

Abstract

Agricultural sector has a significant reliance on insecticides to control insect pests. Since the discovery of the insecticidal properties of dichlorodiphenyltrichloroethane (DDT) in 1939, chemical insecticides have become an integral part of agricultural practices. As the passing dates approach, the demand for these compounds is rapidly increasing. A large amount of chemical insecticides is consumed globally every year. However, the indiscriminate use of these insecticides invites several adverse consequences for the environment and the health of humans and other non-target organisms. In recent days, every entity in the environment might be contaminated with traces of these toxicants, which is of great concern. This situation requires urgent action to protect the existence of life on the Earth's surface. This chapter provides a brief overview of these chemical insecticides, including their historical

---

* Corresponding Author's Email: anokhearchana12@gmail.com; gkdineshiari@gmail.com; isamratsahamtb43@gmail.com

In: Insecticides
Editors: Archana Anokhe and Govindaraj Kamalam Dinesh
ISBN: 979-8-90134-022-6

aspects, the global consumption scenario, and a description of some major classes of chemical insecticides. In the closing portion of this chapter, we have attempted to shed light on why the following chapters are necessary.

Keywords: consequences, environment, global consumption, history, human health, insecticide classes

Introduction

> "Everything is a poison, nothing is a poison. It is the dose that makes the poison."
>
> - Paracelsus

Agriculture is the sector upon which the entire world relies, and the farming community is at the heart of this sector. Agriculture is the backbone of our society, ensuring that people never go to bed hungry. However, the drastic growth in world population is imposing immense pressure on this sector. Perhaps the world population may cross 9.7 billion by 2050 (estimated) (United Nations, 2022), for which expansion in agricultural production is the only way to feed this overgrowing population and achieve Sustainable Development Goal 2: End hunger (United Nations, 2015). To fulfill this need, agricultural production is steadily increasing through the use of advanced technologies (Khan et al., 2021; Hemathilake & Gunathilake, 2022). However, in parallel with agricultural production, the incidence of various pests is also increasing steadily, posing a significant threat to food security. Sarcastically, it can be said that the bond between the pests and crops is a relation made in heaven. The extent of interference humans can cause will never be sufficient to break this bond; in other words, it is never possible to eradicate the pests. Humans can adopt several measures to reduce the severity of these pests and protect crops to a greater extent. Of these measures, the use of chemical pesticides is the most common, widely adopted, and efficient strategy. Pesticides are considered a vital part of crop production due to their adoption in the production of approximately one-third of the global agricultural products (Liu et al., 2002). Pesticides have great potential in controlling pests; therefore, their use in agricultural and horticultural endeavors has yielded tremendous success (Aktar et al., 2009).

Among different groups of pesticides, insecticides contribute significantly by killing or preventing insects and inhibiting their destructive behavior. With the evolution of agriculture, insecticides have become an indispensable tool for crop production and protection sectors (Barathi et al., 2024). Since the 20th century, several chemical insecticides have been introduced to the market, demonstrating promising effects in controlling various insect pests. However, akin to the duality of a coin, the application of insecticides also has inevitable adverse consequences, specifically when these chemicals have been used indiscriminately. Intentionally or unintentionally, extreme reliance on these chemicals leads to environmental (soil, water, and air) and health consequences, as they target non-target ecosystems (Devine & Furlong, 2007; Ren et al., 2023; Barathi et al., 2024). In this chapter, we provide a brief overview of chemical insecticides that may be required before proceeding to the upcoming chapters.

Insecticide Discovery and Evolution: A Historical Narrative That Initiates a New Epoch

Insecticides in the Ancient World and the Pre-DDT Era

Are we still using insecticides from a century ago? Not. The association of insecticides with human civilization dates back to ancient times, albeit with limited usage. A clear description of a catalog of measures to expel fleas from the residence is found in the ancient Egyptian medical papyrus, "Ebers Papyrus," dating back to approximately 1550 BC. In one of the oldest available literary works, the Greek Poet Homer (about 1000 BC) mentioned that the mythological hero Odysseus burned sulfur ".... to purge the hall and the house and the court." The great Greek physician and also the "father of pharmacognosy," Pedanius Dioscorides (40 to 90 AD), mentioned his awareness about the toxic nature of arsenic and sulfur. Records of using arsenic sulfides to control garden insects in China date back as early as 900 AD (Costa, 1987; Freedman, 1995). Hydrocyanic acid was used as a poison by the Egyptians and Romans to fumigate insect pests in museums as early as 1877 (Shepard, 1939). Since 1854, carbon disulfide has been utilized to fumigate insects (Costa, 1987). However, the use of these inorganic compounds was minimal, as insect control was largely reliant on picking or washing off pests until the mid-1800s (Casida & Quistad, 1998). Then came the botanical preparations that gained popularity among the chemists and biologists. For instance, Dalmatian pyrethrum flowers, which contain approximately 1.5% of the active insecticidal ingredient pyrethrin, were widely used as an insecticide in ancient China and subsequently in medieval Persia (Davies et al., 2008), serving as the most crucial botanical for nearly two centuries (Casida & Quistad, 1998). In Europe, dried, ground pyrethrum was sold by Armenian merchants under various names, including "Persian powder," "Persian dust," or "insect powder," and used to control domestic insect pests such as bedbugs, cockroaches, flies, and mosquitoes (Davies et al., 2008). Azadirachtin, rotenone, ryanodine, nicotine, and veratridine were the other botanicals with limited use in this earlier timeframe (Casida & Quistad, 1998). These were the brief histories of insecticides up to the first half of the mid-19th century, when insecticides were mainly of inorganic or botanical origin.

The Discovery of DDT and the Golden Era of Insecticides

Thereafter, the focus has shifted to developing synthetic organic insecticides with enhanced insecticidal properties. In 1874, the Austrian student and chemist Othmar Zeidler synthesized the well-known chemical insecticide dichlorodiphenyltrichloroethane, also known as DDT (Bate, 2007). This breakthrough invention revolutionized the insect pest management sector. Then came the 1930s, a period when the insecticide scenario gained significant momentum with the initiation of the golden era of insecticides. It was in 1939 that the Swiss chemist Paul Müller, in search of an effective insecticide, discovered the insecticidal properties of DDT while working at J. R. Geigy Ltd. in Basel, Switzerland (Oberemok et al., 2015). The practical implications of DDT's performance were beyond belief, leading to statements such as "a single treatment protects crops for a season," "some pests will become extinct," and "mosquito-transmitted diseases will disappear" (Casida & Quistad, 1998). This discovery earned Müller the prestigious Nobel Prize in Physiology or Medicine in 1948. Although then, hardly anyone could believe that such a Nobel Prize-winning discovery would become a devil for the environment in the near future.

The fantastic performance of DDT in the field of insect control led scientists to seek other chlorinated compounds to develop new, cost-effective, and persistent contact insecticides. That is why several chlorinated hydrocarbons, such as hexachlorocyclohexane (HCH), were synthesized initially by Michael Faraday in 1825. The insecticidal property of its γ-isomer was discovered in 1942 (Metcalf, 1955), as well as that of camphene, aldrin, chlordane, and dieldrin, which were introduced into the market quickly after DDT. Finally, this discovery era of chlorinated hydrocarbons came to an end with the development of endosulfan in 1956. The popularity of these chlorinated compounds was so high among the people that within a 40-year window, over 8,000,000,000 pounds of such chlorinated compounds were used (Casida & Quistad, 1998; Li, 1999).

The era of DDT ended with the introduction and widespread use of the organophosphate and carbamate groups of insecticides (Oberemok et al., 2015). While working at Farbenfabriken Bayer, the German chemist Gerhard Schrader discovered the first ever insecticidal organophosphate compound (OPs) in 1937. However, this compound has not been widely used. Thereafter, within a 20-year timeframe, TEPP (Tetraethyl Pyrophosphate), parathion, malathion, diazinon, and many other organophosphate (OP) compounds have been developed. After OPs, carbamate insecticides were developed during testing of physostigmine, a botanical alkaloid derived from the calabar bean. Carbaryl was the first carbamate compound and was very effective against multiple insect pests (Casida & Quistad, 1998). These organophosphate and carbamate compounds are still widely used and play a significant role in managing both domestic and agricultural insect pests.

Silent Spring and the Period of Human Awareness

Until the late 1950s, it was thought that these chemical insecticidal compounds posed little long-term risk to animal health and the environment. However, a rapid increase in the consumption of these compounds has increased public concern regarding their safety status (Casida & Quistad, 1998). A significant event occurred that brought all the people together and changed the scenario of the entire pesticidal industry. In 1962, the American marine biologist and nature writer Rachel Louise Carson, in her ageless book "Silent Spring," criticized the field of applied entomology for applying highly advanced and hazardous chemical pesticides by stating that "It is our alarming misfortune that so primitive a science has armed itself with the most modern and terrible weapons and that in turning them against the insects it has also turned them against the earth." (Carson, 1962). Although her arguments and emotional metaphors were particularly targeted at DDT, the impact of these statements was sensed throughout the entire agrochemical sector, ushering in a new age with stigmatization of the term "pesticide." As a result of public pressure, various regulatory agencies, including the Environmental Protection Agency (EPA), intervened and imposed a ban on DDT in 1972 (Casida & Quistad, 1998). Apart from DDT, Carson was also concerned about the impact of other pesticides, such as other chlorinated hydrocarbons and organophosphates (Davis, 2019).

The Modern Era of Newer Insecticides

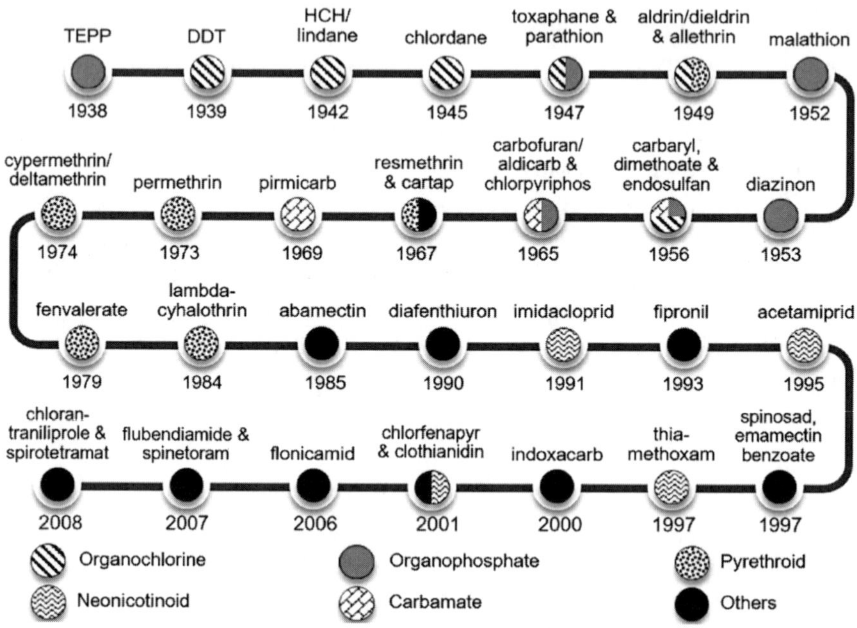

Figure 1. Timeline of introduction or discovery of insecticides from different classes.

Such people's movements forced the agricultural sector to look for alternative, safer chemicals. Therefore, insecticides with lower persistence and safety towards non-target organisms started to capture the market. In this respect, pyrethroids and neonicotinoids, two groups of insecticides, have become dominant in the market. Although the first pyrethroid, allethrin, was synthesized in 1949, the era of pyrethroids gained momentum in the early 1970s with the development of cypermethrin, permethrin, and deltamethrin. These pyrethroid compounds showed low toxicity to warm-blooded animals and did not accumulate in the environment. These features made pyrethroids popular among individuals concerned about their health (Oberemok et al., 2015). The 1990s saw the introduction of neonicotinoids, which partially replaced earlier toxic substances and have become the most widely used class of insecticides worldwide. The first neonicotinoid compound, i.e., imidacloprid, was commercialized in 1991, followed by acetamiprid, clothianidin, and thiamethoxam (Simon-Delso et al., 2015). The search for new insecticides and the improvement of existing ones continues. This push gave rise to newer classes of insecticides, such as macrocyclic lactones, phenylpyrazoles, thioureas, nereis toxin analogues, spinosyns, oxadiazines, diamides, benzoylureas, and cyclic ketoenols, with novel modes of action and higher efficacy against multiple insect pests (Oberemok et al., 2015). However, in recent years, with technological advancements and increasing human concerns about the health consequences of chemical insecticides, more focus has shifted towards safer options, such as microbial biopesticides, biochemical pesticides, and plant-incorporated protectants (PIPs). This is a summary of the tumultuous history of insecticides. In Figure 1, we have presented a chronology of the discovery of insecticides.

Global Status of Insecticide Consumption

Insecticides or any other forms of chemical pesticides play a very crucial role in both agricultural, horticultural, and domestic pest control sectors. In addition, the last few decades have shown rapid growth in the world population, which in turn causes a sharp increase in demand for chemical pesticides. That is why a closer look will ensure that we are treating our food plants more frequently with a greater amount of pesticides than ever before. However, there is a lack of data solely on insecticides, as most data typically represent all types of pesticides. That is why a brief global scenario of pesticides has been documented in this section.

Between 2000 and 2021, the global usage of pesticides increased by 62%. The majority of this increase occurred between 2000 and 2016, with a brief dip until 2018, followed by another phase of expansion thereafter (FAO, 2023). Approximately 4.2 million metric tons (mmt) of pesticides are consumed globally each year. They are expected to reach 4.6 mmt by 2026 (Report Linker, 2023). According to the Food and Agriculture Organization (FAO) report, in 2020, approximately 1.8 kg of pesticide was used per hectare of cropland worldwide, with Saint Lucia (20.48 kg/ha) applying the highest amount of pesticide per hectare of cropland (FAO, 2022). Figure 2 illustrates the global scenario of pesticide consumption per cropland area, based on FAO 2022 statistics. According to recent data, China is the largest user of pesticides, followed by the USA, Brazil, Argentina, Canada, Ukraine, France, Malaysia, Australia, and Spain (Worldometer, 2023). The current annual import of pesticides is 2.5 million metric tons (mmt) and is expected to reach 2.8 mmt by 2026, following an annual growth rate of 2.4% since 2016 (ReportLinker, 2023). According to data from the World Health Organization (WHO), developing countries worldwide consume ~20% of pesticides, with an increasing growth rate (Sharma et al., 2019). Moreover, in low-income countries, global pesticide consumption increased by 153% in volume over the last decade (Shattuck et al., 2023). Regarding the continent-wise share in global pesticide consumption, African countries contribute only 2–4% of the total global pesticide market share and have the lowest rate of pesticide usage, despite their population of 59% being dependent on farming (Abate et al., 2000; FAO, 2022). Conversely, the American continent leads in both pesticide consumption (~1.78 million metric tons in 2021) and the rate of pesticide application (~4.7 kilograms per hectare in 2021). The share of the United States in worldwide pesticide consumption increased from 41% to 50%. In contrast, Asian and European countries experienced a 5-6% reduction in pesticide consumption (FAO, 2023). Although the application of pesticides in the agricultural sector is increasing rapidly in Southeast Asian countries (Kunstadter, 2007), the usage of pesticides is still relatively low in Asian countries, with an application rate of 1.17 kg/ha, even lower than the global average (FAO, 2022).

Now, regarding the insecticide, its contribution among different classes of pesticides used globally is relatively low, with a share of only 18%. In contrast, herbicides alone contribute approximately 52% share, followed by fungicides (23%). The rest are contributed by other forms of pesticides, such as nematicides and rodenticides (FAO, 2022). However, even with a smaller share, the insecticide market remains a substantial one, valued at 14 billion USD in 2021 and projected to reach 22 billion USD by 2030, with a compound annual growth rate (CAGR) of 5% (Straitsresearch, 2023). Specifically, the Asian population uses a greater amount of insecticides compared to other parts of the world. Conversely, in European countries, insecticides contribute only 12% of their total pesticide use (FAO, 2022). According to data, between 1990 and 2010, 342,000 metric tons of insecticides were consumed annually by 82 nations, with the USA contributing the most significant amount, at 90,000 metric tons per year, followed by India (Liu et al., 2015). Among different classes of insecticides, the neonicotinoid group holds a considerable market share. In 1990, the global market for insecticides was dominated by organophosphates, carbamates, and synthetic pyrethroids. However, since entering the global insecticide market in 1991, neonicotinoids have surpassed all others in terms of global usage, covering a significant share of the global insecticide market (Jeschke et al., 2011). Once the patent protections were lifted in the mid-2000s, sales of neonicotinoid insecticides skyrocketed with an increase of 1.6 to 14.6 times in the sales of individual neonicotinoid insecticides between 2003 and 2009 (Simon-Delso et al., 2015). By 2014, their global insecticide market share reached up to 25% (Bass et al., 2015). Even by 2008, imidacloprid had emerged as the highest-selling insecticide in the world and the second-highest-selling pesticide after glyphosate, with registered usage in 140 crops across 120 countries (Jeschke et al., 2011).

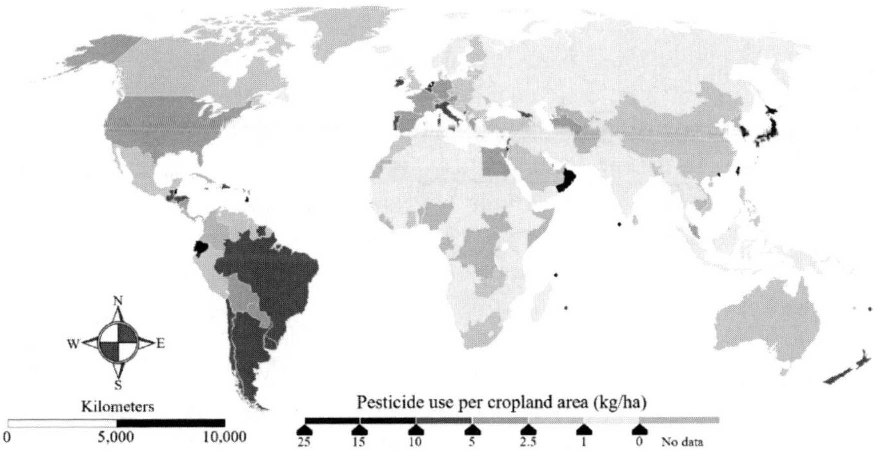

Source: https://www.fao.org/faostat/en/#data/EP.

Figure 2. Global scenario of pesticide consumption per cropland area during 2020.

However, the consumption of insecticides in many countries is now declining due to the adoption of biopesticides, genetically modified crops, and other natural plant products for insect pest management (Sharma et al., 2019; Fusar Poli & Fontefrancesco, 2024). According to recent data, some global leaders in the insecticide market include Syngenta, Bayer CropScience, BASF, Corteva, UPL, Adama, FMC, and Sumitomo, which collectively hold the largest market share (Statista, 2024). Our country, India, follows a distinct pattern of pesticide use compared to other nations, with insecticides accounting for the significant share of pesticides (62.2%), followed by fungicides (19.2%), herbicides (14.4%), and other types (4.2%). With over 219 manufacturing facilities and more than 4,000 formulation facilities, India has a substantial pesticide production capacity (Barathi et al., 2024). India also holds the 12th rank globally in terms of insecticide usage. Among Asian countries, India is the second-largest producer of insecticides after China (Shahid & Khan, 2022). However, the per-hectare consumption of insecticides on cropland is much lower in India.

A Short Description of Some Important Classes of Insecticides

As stated before, insecticides have a long history. A wide range of natural organic insecticides, including plant-derived phytochemicals like terpenes, alkaloids, saponins, phenolic compounds, etc. with insecticidal properties (Boulogne et al., 2012; Tlak Gajger & Dar, 2021); plant-derived essential oil (Ikbal & Pavela, 2019; Isman, 2020); oilseed cakes (Rao et al., 2012); etc., have been proven to be active against a variety of insect pests with reduced environmental persistency, low toxicity to mammals, and have complex chemistry that can avoid insect pest resistance issues (George et al., 2014). Inorganic insecticides, such as mercury, arsenicals, and sulfur, have a long history of insecticidal use (Norton, 1948). Furthermore, in recent days, there have been high demands for biopesticides and genetically engineered crops (Damalas & Koutroubas, 2018; Kumar et al., 2020). However, among different categories of insecticides, the most widely used and in-demand category is synthetic insecticides. Although the era of synthetic insecticides began with the discovery of DDT, a representative of the organochlorine class of insecticides, the occurrence of resistance in insects and an increase in human concern necessitated the development of new insecticides with different modes of action. Currently, several synthetic insecticide classes are available on the market. Among them, organochlorine, carbamate, organophosphate, pyrethroid, and neonicotinoid are the most widely used classes of synthetic insecticides, each having different modes of action against insect pests (Bilal et al., 2019). In this section, a brief description of these classes of insecticides is provided.

Organochlorines

Organochlorine or chlorinated hydrocarbon insecticides are organic chemicals containing five or more chlorine atoms. This group constitutes one of the initial representatives of insecticides ever developed and is employed in agriculture. Organochlorine insecticides can be categorized into three basic sub-groups that include dichlorodiphenylethanes (e.g., DDT, methoxychlor, dicofol, etc.), chlorinated benzenes and cyclohexanes (e.g., HCH, chlordecone, lindane, mirex, toxaphene, etc.), and chlorinated cyclodienes (e.g., aldrin, dieldrin, endrin, chlordane, endosulfan, heptachlor, etc.) (Tsai, 2024). These insecticides enter the animal body either orally or topically, as absorption of these compounds is rapid due to their lipid-soluble nature, and sometimes enter through inhalation (Marth et al., 1989). Organochlorines exhibit greater efficiency in killing insects, as evidenced by lower LD_{50} (median lethal dose) values (Kaushik & Kaushik, 2007). These insecticides work through two mechanisms. The first one is for DDT-type organochlorines that affect the brain and peripheral nervous system by slowing down the process of Na^+ (sodium) influx and inhibiting K^+ (potassium) efflux, resulting in an excess accumulation of intracellular K^+. This depolarizes the cell partially. The threshold for an additional action potential is lowered, leading to premature depolarization of the neuron. This leads to the tremors and other nerve malfunctions, a symptom known as "DDT jitters" (Davies et al., 2008; Smith, 2012; Ensley, 2018). The second one is for the cyclodiene insecticides, which function by competitively inhibiting GABA (gamma-aminobutyric acid) binding at its receptor. This GABA-gated chloride channel induces chloride ion uptake in neurons, thereby calming nerve transmission. Inhibition of this GABA activity by cyclodiene poisoning causes partial neuronal repolarization, resulting in uncontrolled excitation of neurons (Narahashi et al., 1992; Kaushik & Kaushik, 2007).

This class of synthetic insecticides poses a serious threat to the environment and human health due to the high persistence of these toxicants in the environment, their ability to transport between phases, and their capacity to bioaccumulate in the food web (Tsai, 2024). That is why, once the most commonly used class of insecticide in the world, organochlorine is composed of representatives, the majority of which are now banned in most of the developed countries (Ensley, 2018; Keswani et al., 2022; Khuman et al., 2022). However, specific toxicants of this group are still produced and applied in the vector control programmes in many tropical developing countries

(Abubakar et al., 2020). Even after this ban, these chemicals continue to cause adverse effects on animal health worldwide (Ensley, 2018).

Organophosphates

After organochlorines, another group of insecticides that gained considerable popularity among farmers was organophosphates (OPs). Since World War II, several thousand OP compounds have been synthesized for various purposes, making them one of the largest classes of chemicals. These compounds are derived from phosphoric acid, carrying phosphoryl (P=O) or thiophosphoryl (P = S) groups. These compounds have a chemically and thermally inert C–P direct linkage that makes them resistant to extreme situations, such as chemical hydrolysis, photolysis, and thermal breakdown, compared to analogous compounds with more reactive O–P, N–P, or S–P linkages (Abubakar et al., 2020). OP compounds have complex chemistry, primarily due to the type and position of various side chains attached to the phosphorus atom, which creates confusion in their classification (Gupta, 2006).

Most of these compounds possess pesticidal properties, while others are utilized as flame retardants, nerve agents, and parasiticides in veterinary medicine. This heterogeneous group of toxicants has a broad target spectrum, effective against a wide range of insect pests (Gupta, 2006). The OP compounds can be divided into three basic sub-groups, namely aliphatic group (dichlorvos, dimethoate, malathion, monocrotophos, acephate, methamidophos, TEPP, etc.), phenyl group (profenofos, fenthion, methyl parathion, etc.), and heterocyclic derivatives group (chlorpyrifos, diazinon, etc.) (Reddy, 2018). Typically, these compounds act by inhibiting the acetylcholine esterase (AChE) enzyme. This substance breaks down the neurotransmitter acetylcholine into acetic acid and choline, thereby stopping the nerve impulse. That is why inhibition of this enzymatic function causes overstimulation of the nervous system, leading to rapid twitching of muscles, paralysis, and ultimately, death (Main, 1979). OP chemicals have two distinct features. The first is their biodegradability or non-persistence in nature, which causes less environmental pollution. Secondly, these chemicals possess higher toxicity towards vertebrates (Reddy, 2018). Still, the contribution of this group of chemicals is significant in improving agricultural productivity and effective crop yields (Abubakar et al., 2020).

Carbamates

Carbamates are organic insecticides derived from carbamic acid, with the general formula ROC(O)NHCH3, and possess both contact and stomach action. Unlike organophosphates, carbamates have a much simpler structure (Gupta, 2006). In the case of most compounds, R denotes a substituted phenol or heterocycle, with exceptions like aldicarb and pirimicarb. Generally, carbamates exhibit poor solubility in water but show increased solubility in polar organic solvents, such as ethanol, methanol, and acetone (Vale & Lotti, 2015). Carbamates can be sub-grouped into heterocyclic (isolan), naphthyl (carbaryl), phenyl (carbofuran), and oxime (aldicarb) carbamates (Prasad, 2019). Like OPs, carbamates also act by inhibiting AChE. The inhibition occurred due to the carbamylation of a serine hydroxyl residue at the enzyme's active site, a process involving the breakdown of the carbamate molecule, which AChE interprets as acetylcholine (Alvares et al., 1992). However, unlike OPs, the inhibition achieved by carbamate insecticides is reversible, specifically due to the spontaneous hydrolysis of the carbamylated enzyme that causes AChE to restore its activity (Vale & Lotti, 2015). Carbamate compounds are also widely used for controlling insect pests. Some important carbamate insecticides include carbaryl, carbofuran, carbosulfan, aldicarb, oxamyl, methomyl, pirimicarb, propoxur (against household insects like cockroaches), etc. In addition to their use as insecticides, carbamates are also employed as drugs in human medicine to treat Alzheimer's disease, glaucoma, and myasthenia gravis, as well as in veterinary medicine to control parasites (Sugimoto et al., 2002; Gupta, 2006). Carbamate insecticides also possess systemic toxicity upon inhalation, ingestion, or skin contamination (Sarkar et al., 1994). However, carbamate poisoning is considered to be less severe than organophosphate poisoning (Vale & Lotti, 2015).

Synthetic Pyrethroids

Another group of widely used insecticides is synthetic pyrethroids. The naturally occurring extract, pyrethrum, is isolated from the flowers of ornamental plants *Chrysanthemum cinerariifolium* and *Chrysanthemum coccineum*. Over an extended period, extracts of pyrethrum and, more recently, various synthetic chemical analogs of pyrethrum have been produced and widely used for insect pest control. Pyrethrum consists of six active insecticidal substances, including pyrethrin I, pyrethrin II, cinerin I, cinerin II, jasmolin I, and jasmolin II, of which pyrethrin I and pyrethrin II are the major active ones (Rehman et al., 2014). All these compounds together are known as pyrethrins. Synthetic pyrethroids are nothing but the synthetic analogs and derivatives of the pyrethrins and encompass a diverse category of about 1,000 insecticides. Even after being analogs of natural pyrethrins, synthetic pyrethroid production undergoes extensive chemical changes, making these compounds highly toxic and more persistent in the environment than natural pyrethrins (Thatheyus & Selvam, 2013). Some of the most commonly used synthetic pyrethroids in the agricultural sector include cypermethrin, permethrin, deltamethrin, fenevelerate, λ-cyhalothrin, α-cypermethrin, etc. (Abubakar et al., 2020). Synthetic pyrethroids exhibit modes of action similar to those of DDT. They primarily act on sodium and chloride channels, lowering the action potential threshold of the nervous system and causing repeated stimulation and paralysis (Narahashi et al., 1998; Aznar-Alemany & Eljarrat, 2020). Synthetic pyrethroids possess high neurotoxic activity to insects and fish and are quite safer to mammals and birds.

Furthermore, these toxicants are not photostable and break quickly in the presence of light. That is why this insecticide group is considered to be safer for use in food production (Abubakar et al., 2020). However, with technological advancements, the photostability of these compounds has increased. Over the past two decades, synthetic pyrethroids have largely supplanted organophosphate insecticides in most of their applications, much like organophosphates had replaced organochlorinated insecticides earlier (Aznar-Alemany & Eljarrat, 2020).

Neonicotinoids

Currently, the most widely used group of chemical insecticides is neonicotinoids. Since the 2000s, the usage of neonicotinoids has rapidly increased due to the alterations in application methodologies brought forth by the increased usage of coated seeds, the rise in insect resistance issues, and/or worries about the high toxicity of other previously used insecticides (organochlorines, organophosphates, and carbamates) to mammals (Jeschke et al., 2011; Douglas & Tooker, 2015). Neonicotinoids have a structure similar to natural nicotine. Neonicotinoids can be subdivided into three chemical sub-groups, including the N-nitroguanidines (clothianidin, dinotefuran, imidacloprid, and thiamethoxam), N-cyanoamidines (acetamiprid and thiacloprid), and nitromethylenes (nitenpyram) (Goulson, 2013).

Neonicotinoid compounds have good water solubility (log K_{ow}: 0.55–1.26; log K_{oc}: 1.4−2.3) with somewhat persistency in aquatic and soil environments and can get transported to a distance away from the area of initial application (Hladik et al., 2018). These insecticides exhibit good systemic action, rapidly being taken up by plants and transported through all plant tissues, which provides particular advantages in insect pest management, particularly for managing boring and root-feeding insects, against which non-systemic compounds are ineffective (Goulson, 2013). Neonicotinoids act as nicotinic acetylcholine receptor (nAChR) agonists. Binding of these compounds to nAChRs in the insect's central nervous system causes rapid nerve stimulation, followed by paralysis and death at higher insecticide concentrations. However, this binding is stronger to the nAChRs of insects than to those of vertebrates, making these compounds more selective for insects and safer for vertebrates (Tomizawa & Casida, 2003, 2005; Goulson, 2013; Simon-Delso et al., 2015). One of the significant uses of neonicotinoids in agriculture is as seed coatings. In addition, these compounds are also used as foliar sprays, in-furrow treatments, and granules (Hladik et al., 2018). Some of the most commonly used neonicotinoids in the agricultural sector include imidacloprid, acetamiprid, thiamethoxam, clothianidin, and dinotefuran. Although neonicotinoids contributed significantly to agricultural growth, their toxicity to some beneficial insects is of great concern. Specifically, regarding honey bees, these compounds are considered a major contributor to colony collapse disorder (CCD) (Lu et al., 2014, 2020), which has drawn attention to the environmental impact of these compounds.

Others

Apart from these five major classes of insecticides, some other groups have also been widely used in agriculture. Of them, nereistoxin analogues (cartap, bensultap, thiocyclam), thiourea compound (diafenthioron), phenylpyrazole (fipronil), diamide (flubendiamide, chlorantraniliprole, cyantraniliprole), oxadiazine (indoxacarb), avermectins, spinosad, halogenated pyrroles (chlorfenapyr), benzoylureas (diflubenzuron, lufenuron, novaluron), etc., are the important ones. More recently, the application of these chemical compounds has increased due to the lower toxicity of some of these compounds and their higher effectiveness compared to earlier groups of insecticides.

In Figure 3, the chemical structures of some important insecticides from different classes used in the agricultural sector are presented.

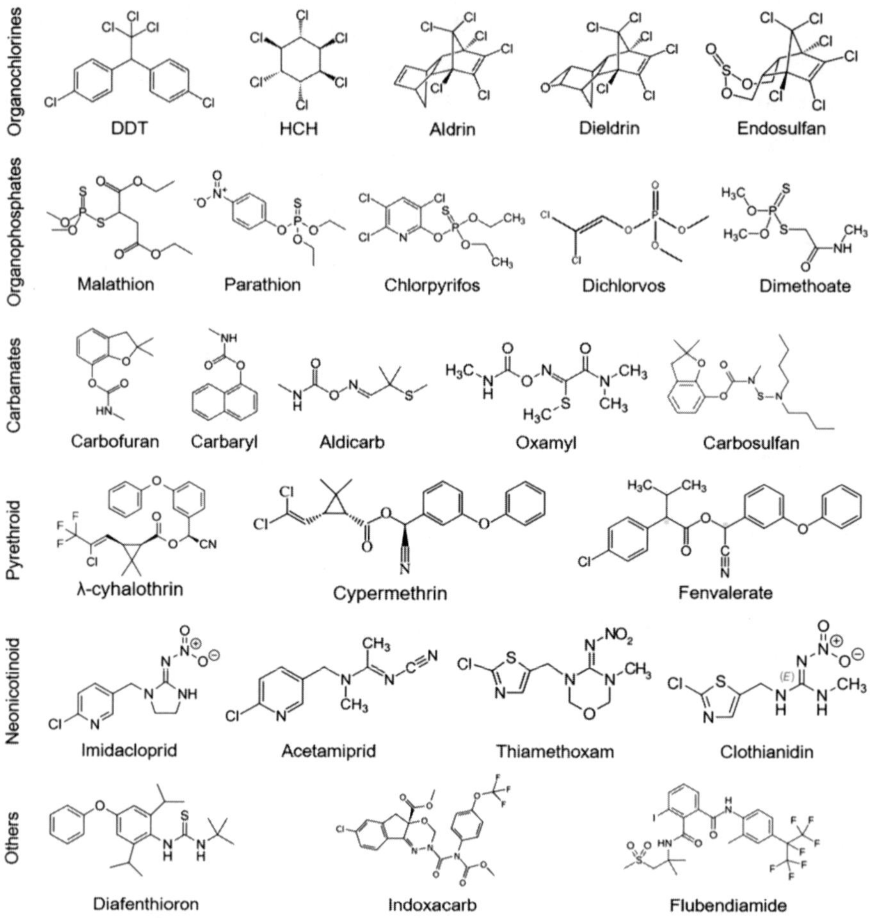

Figure 3. Chemical structure of some commonly used synthetic insecticides.

Insecticides vs the Environment: Things That We Need to Know

There is no doubt that these chemical insecticides have a significant contribution to boosting our agricultural productivity. However, we cannot overlook the serious threat that these toxicants pose to various biological entities in the natural world. As a part of nature, we need to know certain things in order to protect the environment and ourselves from these hazardous chemicals. The purpose of this section is to introduce the necessity of the upcoming chapters with a brief overview of their contents.

Consequences of Insecticides on the Environment

These chemical insecticides have deleterious effects on the environment, a fact that is well known to us. However, how these insecticides pose threats to the environment, what these threats are, and which environmental entities are most highly threatened by these toxicants are certain things that we need to be aware of. Almost every part of our environment is highly contaminated with these chemicals. Their remnants and residues have been found in air, soil, sediments, surface and underground water, thus deteriorating their qualities (Ansari et al., 2014; Tang et al., 2018; Zhang et al., 2023). That is why the living entities relying on these abiotic factors are in great danger. The way these chemicals affect soil nutrient recycling, soil fertility and productivity, as well as the soil-associated micro- and macro-flora and fauna, is of great concern. It is also necessary for us to know the impact of these toxic chemicals on the aquatic system, which is the home for many aquatic flora and fauna. We also need to be concerned about the worldwide deterioration of the air quality index (AQI) due to volatile organic insecticidal compounds and insecticide drift caused by wind.

Furthermore, the impact of these chemicals on non-target beneficial organisms, such as pollinators, natural enemies, and other animals, is of considerable concern and warrants further attention. The indiscriminate use of chemical insecticides also causes phenomena such as pest resistance, resurgence, and secondary pest outbreaks, which directly or indirectly increase the incidence of insect pests. Together, these factors push the environment in an undesirable direction, making it a terrible place to live. As part of this environment, we need to be aware of these circumstances that are knocking on our doors.

Consequences of Insecticides on Human Health

As stated above, insecticides have serious deleterious effects on various components of the environment. Since we are also part of it, the consequences on our health are of extreme importance. Insecticide poisoning may cause both acute and chronic poisoning. Acute poisoning is a prevalent thing. People often do not bother about the effects of insecticides, as they consume or inhale these toxicants at very low concentrations every day. However, accumulation of these toxicants in the human body day after day may cause serious chronic health issues that can be seen years later. Diseases such as asthma, dermatitis, cancer, infertility, endocrine problems, neurobehavioral disorders, Parkinson's and Alzheimer's disease, immunotoxicity, and congenital abnormalities are associated with chronic exposure to these insecticides (Kim et al., 2017; Beyuo et al., 2024). Specifically, for the children, insecticide poisoning has the most detrimental effects due to their weak immune system. Reports of child deaths due to acute insecticide poisoning have been documented in many countries, such as Canada, India, Turkey, South Korea, South Africa, and Tanzania (Lekei et al., 2020; Islam et al., 2022). For farmers, insecticide poisoning is a common occurrence due to their prolonged exposure to insecticides without taking any safety measures. Every year, several people worldwide lose their lives because of the direct or indirect consequences of insecticide poisoning. Nonetheless, people, particularly in developing countries, are unconscious of these issues. However, such consequences can be mitigated through the adoption of simple precautions, proper mitigation strategies, effective insecticide handling techniques, and by raising awareness. Furthermore, different classes of insecticides pose distinct health risks to humans that people are often unaware of. These are certain things that people need to know to safeguard their health from insecticide-driven hazards, which will be elaborated upon in the upcoming chapters.

Insecticide Detection in the Environment

Considering the deleterious impacts of chemical insecticides on both biotic and abiotic ecosystems, detecting insecticides in the environment is crucial. This is a technical issue that the global scientific community and the younger generation should be concerned about. After application, the remnants and residues of these toxicants accumulate in various environmental factors. To reduce the level of risks to our environment and the health of living beings, residues must be within the permissible limits. Therefore, the determination of insecticide residues in various food commodities is a pressing need for the benefit of consumers, producers, and authorities responsible for food quality control (Aguilera et al., 2003). These traces can be detected using various techniques, ranging from simple to complex, that are often beyond the knowledge of most people. Techniques such as various types of chromatography, spectrometry, immunoassays, and electrochemical techniques have been widely accepted globally, particularly in terms of toxicity and environmental safety. Further technological advancements are also introducing new and efficient detection techniques. However, different techniques have their strengths and limitations.

Furthermore, the implications of these techniques vary depending on the nature of the toxicant and various other factors. By adopting these techniques, various laboratories can ensure the detection of insecticide contamination in foods, soil, water, and air, among other substances, in their corresponding localities. For that, knowledge of these techniques is a prerequisite, as demonstrated in the following chapters.

Insecticide Remediation

So, we have already discussed the inevitable consequences of chemical insecticides that we need to be concerned about. Considering the impact of these consequences, their remediation is of extreme importance. The serious threats posed by these chemical insecticides necessitate urgent mitigation strategies. Several tactics, such as various physical (soil vapor extraction, activated carbon adsorption, membrane filtration), chemical (advanced oxidation processes, photocatalysis), biological (microbial degradation, phytoremediation), and nanotechnology-based approaches (nanosensors, nano-remediation), can be adopted to enhance the detection and removal of insecticides at low concentrations. However, the success of these tactics depends on proper knowledge of their implementation. Chapter 5 of this book has covered this crucial issue.

Regulatory and Policy Issues in Insecticide Use

In addition to understanding the consequences of chemical insecticides, people also need to be aware of the regulatory and policy concerns associated with insecticide use. Every hazardous material needs the implementation of proper regulatory actions. With increasing consequences of these toxicants, regulatory and policy issues have become significantly important in recent years. The issues related to the development and implementation of regulations governing insecticide use, the evaluation of hazards associated with consumption, and the development of alternative pest control strategies require further exploration and attention.

These are certain things that we need to know in order to protect the environment and ourselves from these hazardous chemicals. In the upcoming chapters, the concerns that chemical insecticides pose to biotic and abiotic ecosystems are described in detail, enriching readers' knowledge on these critical issues. Furthermore, techniques for detecting insecticides in the environment and ways to lessen their hazards are detailed extensively in the sections that follow. A chapter has also been dedicated to discussing the regulatory and policy issues in insecticide use.

Conclusion

In today's world, chemical insecticides pose a significant threat to the existence of all forms of life on the Earth's surface. Since their development to date, the damage these toxicants have caused to the environment and human health remains unquantifiable. Still, the dependency on these chemicals is increasing, specifically in the agricultural sector. When one chemical fails or is banned, another toxic one quickly takes its place. As demand increases, the threat imposed by them is also snowballing. In this context, the quote by Rachel Carson, "… man is a part of nature, and his war against nature is inevitably a war against himself" (Carson, 1962) is highly justifiable. That is why it is our job to take action against these hazardous compounds to protect ourselves. In recent days, many researchers, social activists, and government and non-government organizations around the world have taken initiatives to tackle these hazardous situations. To address the challenges posed by pesticides, the UN Convention on Biological Diversity adopted the Kunming-Montreal Global Biodiversity Framework in December 2022, with a key target to reduce the risks associated with pesticide pollution by at least half by 2030 (CBD 2022). Similarly, numerous initiatives have been undertaken at both national and international levels. Still, it needs more emphasis from the general public, who are mostly threatened by these toxins. The best way to gather their focus on this issue is by making them aware of the different aspects of these chemical insecticides, specifically their impact on the environment. Ultimately, we are part of "Mother Nature," and only by protecting her can we protect ourselves.

References

Abate, T., van Huis, A., & Ampofo, J. K. O. (2000). Pest management strategies in traditional agriculture: an African perspective. *Annual Review of Entomology*, *45*(1), 631–659. https://doi.org/10.1146/annurev.ento.45.1.631.

Abubakar, Y., Tijjani, H., Egbuna, C., Adetunji, C. O., Kala, S., Kryeziu, T. L., Ifemeje, J. C., & Patrick-Iwuanyanwu, K. C. (2020). Pesticides, history, and classification. In C. Egbuna & B. Sawicka (Eds.), *Natural Remedies for Pest, Disease and Weed Control* (pp. 29–42). Academic Press. https://doi.org/10.1016/B978-0-12-819304-4.00003-8.

Aguilera, A., Brotons, M., Rodríguez, M., & Valverde, A. (2003). Supercritical fluid extraction of pesticides from a table-ready food composite of plant origin (gazpacho). *Journal of Agricultural and food Chemistry*, *51*(19), 5616–5621. https://doi.org/10.1021/jf030233k.

Aktar, M. W., Sengupta, D., & Chowdhury, A. (2009). Impact of pesticides use in agriculture: their benefits and hazards. *Interdisciplinary Toxicology*, *2*(1), 1–12. http://doi.org/10.2478/v10102-009-0001-7.

Alvares, A. P. (1992). Pharmacology and toxicology of carbamates. In B. Ballantyne & T. C. Marrs (Eds.), *Clinical and Experimental Toxicology of Organophosphates and Carbamates* (pp. 40–46). https://doi.org/10.1016/B978-0-7506-0271-6.50011-5.

Ansari, M. S., Moraiet, M. A., & Ahmad, S. (2014). Insecticides: impact on the environment and human health. In A. Malik, E. Grohmann, & R. Akhtar (Eds.), *Environmental Deterioration and Human Health: Natural and Anthropogenic Determinants* (pp. 99–123), Springer Dordrecht. https://doi.org/10.1007/978-94-007-7890-0_6.

Aznar-Alemany, Ò., & Eljarrat, E. (2020). Introduction to pyrethroid insecticides: Chemical structures, properties, mode of action and use. In E. Eljarrat (Ed.), *Pyrethroid Insecticides* (pp. 1–16). Springer Cham. https://doi.org/10.1007/698_2019_435.

Barathi, S., Sabapathi, N., Kandasamy, S., & Lee, J. (2024). Present status of insecticide impacts and eco-friendly approaches for remediation-a review. *Environmental Research*, *240*, 117432. https://doi.org/10.1016/j.envres.2023.117432.

Bass, C., Denholm, I., Williamson, M. S., & Nauen, R. (2015). The global status of insect resistance to neonicotinoid insecticides. *Pesticide Biochemistry and Physiology*, *121*, 78–87. https://doi.org/10.1016/j.pestbp.2015.04.004.

Bate R. (2007). The rise, fall, rise, and imminent fall of DDT. *American Enterprise Institute for Public Policy Research* 14(4), 1–9.

Beyuo, J., Sackey, L. N., Yeboah, C., Kayoung, P. Y., & Koudadje, D. (2024). The implications of pesticide residue in food crops on human health: a critical review. *Discover Agriculture*, *2*(1), 123. https://doi.org/10.1007/s44279-024-00141-z.

Bilal, M., Iqbal, H. M., & Barceló, D. (2019). Persistence of pesticides-based contaminants in the environment and their effective degradation using laccase-assisted biocatalytic systems. *Science of The Total Environment*, *695*, 133896. https://doi.org/10.1016/j.scitotenv.2019.133896.

Boulogne, I., Petit, P., Ozier-Lafontaine, H., Desfontaines, L., & Loranger-Merciris, G. (2012). Insecticidal and antifungal chemicals produced by plants: a review.

Environmental Chemistry Letters, *10*(4), 325–347. https://doi.org/10.1007/s10311-012-0359-1.

Carson, R. (1962). *Silent Spring*, Houghton Mifflin.

Casida, J. E., & Quistad, G. B. (1998). Golden age of insecticide research: past, present, or future?. *Annual Review of Entomology*, *43*(1), 1–16. https://doi.org/10.1146/annurev.ento.43.1.1.

CBD. 2022. Kunming-Montreal Global biodiversity framework. Draft decision submitted by the President. CBD/COP/15/L.25 Conference of the Parties to the Convention on Biological Diversity. Fifteenth meeting – Part II. Montreal, Canada, 7-19 December 2022. https://www.cbd.int/doc/c/e6d3/cd1d/daf663719a03902a9b116c34/cop-15-l-25-en.pdf.

Costa, L. G. (1987). Toxicology of pesticides: A brief history. In L. G. Costa, C. L. Galli, & S. D. Murphy (Eds.), *Toxicology of Pesticides: Experimental, Clinical and Regulatory Perspectives* (pp. 1–10). Springer, Berlin, Heidelberg. https://doi.org/10.1007/978-3-642-70898-5_1.

Damalas, C. A., & Koutroubas, S. D. (2018). Current status and recent developments in biopesticide use. *Agriculture*, *8*(1), 13. https://doi.org/10.3390/agriculture8010013.

Davies, T. G. E., Field, L. M., Usherwood, P. N. R., & Williamson, M. S. (2008). DDT, pyrethrins, pyrethroids and insect sodium channels. *IUBMB Life*, *59*(3), 151–162. https://doi.org/10.1080/15216540701352042.

Davis, F. R. (2019). Pesticides and the perils of synecdoche in the history of science and environmental history. *History of Science*, *57*(4), 469–492. https://doi.org/10.1177/0073275319848964.

Devine, G. J., & Furlong, M. J. (2007). Insecticide use: Contexts and ecological consequences. *Agriculture and Human Values*, *24*, 281–306. https://doi.org/10.1007/s10460-007-9067-z.

Douglas, M. R., & Tooker, J. F. (2015). Large-scale deployment of seed treatments has driven rapid increase in use of neonicotinoid insecticides and preemptive pest management in US field crops. *Environmental Science & Technology*, *49*(8), 5088–5097. https://doi.org/10.1021/es506141g.

Ensley, S. M. (2018). Organochlorines. In R. C. Gupta (Ed.), *Veterinary Toxicology: Basic and Clinical Principles*. (pp. 509–513). Academic Press. https://doi.org/10.1016/B978-0-12-811410-0.00038-6.

FAO. (2022). *Pesticides use, pesticides trade and pesticides indicators: Global, regional and country trends, 1990–2020* (FAOSTAT analytical brief 16). Food and Agriculture Organization of the United Nations, Rome.

FAO. (2023). *World Food and Agriculture – Statistical Yearbook 2023*. Rome. https://doi.org/10.4060/cc8166en.

Freedman, B. (1995). Pesticides. In T*he Ecological Effects of Pollution, Disturbance, and Other Stresses* (2nd ed., pp. 213–277). Academic Press. https://doi.org/10.1016/B978-0-08-050577-0.50013-3.

Fusar Poli, E., & Fontefrancesco, M. F. (2024). Trends in the implementation of biopesticides in the Euro-Mediterranean region: A narrative literary review. *Sustainable Earth Reviews*, *7*(1), 14. https://doi.org/10.1186/s42055-024-00085-8.

George, D. R., Finn, R. D., Graham, K. M., & Sparagano, O. A. (2014). Present and future potential of plant-derived products to control arthropods of veterinary and medical significance. *Parasites & vectors*, *7*, 28. https://doi.org/10.1186/1756-3305-7-28.

Goulson, D. (2013). An overview of the environmental risks posed by neonicotinoid insecticides. *Journal of Applied Ecology*, *50*(4), 977–987. https://doi.org/10.1111/1365-2664.12111.

Gupta, R. C. (2006). Classification and uses of organophosphates and carbamates. In *Toxicology of Organophosphate & Carbamate Compounds* (pp. 5–24). Academic Press. https://doi.org/10.1016/B978-012088523-7/50003-X.

Hemathilake, D. M. K. S., & Gunathilake, D. M. C. C. (2022). Agricultural productivity and food supply to meet increased demands. In R. Bhat (Ed.), *Future Foods: Global Trends, Opportunities, and Sustainability Challenges* (pp. 539–553). Academic Press. https://doi.org/10.1016/B978-0-323-91001-9.00016-5.

Hladik, M. L., Main, A. R., & Goulson, D. (2018). Environmental risks and challenges associated with neonicotinoid insecticides. *Environmental Science & Technology*, *52*(6), 3329–3335. https://doi.org/10.1021/acs.est.7b06388.

Ikbal, C., & Pavela, R. (2019). Essential oils as active ingredients of botanical insecticides against aphids. *Journal of Pest Science*, *92*, 971–986. https://doi.org/10.1007/s10340-019-01089-6.

Islam, M. A., Amin, S. N., Rahman, M. A., Juraimi, A. S., Uddin, M. K., Brown, C. L., & Arshad, A. (2022). Chronic effects of organic pesticides on the aquatic environment and human health: A review. *Environmental Nanotechnology, Monitoring & Management*, *18*, 100740. https://doi.org/10.1016/j.enmm.2022.100740.

Isman, M. B. (2020). Commercial development of plant essential oils and their constituents as active ingredients in bioinsecticides. *Phytochemistry Reviews*, *19*, 235–241. https://doi.org/10.1007/s11101-019-09653-9.

Jeschke, P., Nauen, R., Schindler, M., & Elbert, A. (2011). Overview of the status and global strategy for neonicotinoids. *Journal of Agricultural and Food Chemistry*, *59*(7), 2897-2908. https://doi.org/10.1021/jf101303g.

Kaushik, P., & Kaushik, G. (2007). An assessment of structure and toxicity correlation in organochlorine pesticides. *Journal of Hazardous Materials*, *143*(1-2), 102–111. https://doi.org/10.1016/j.jhazmat.2006.08.073.

Keswani, C., Dilnashin, H., Birla, H., Roy, P., Tyagi, R. K., Singh, D., Rajput, V. D., Minkina, T., & Singh, S. P. (2022). Global footprints of organochlorine pesticides: A pan-global survey. *Environmental Geochemistry and Health*, *44*, 149–177. https://doi.org/10.1007/s10653-021-00946-7.

Khan, N., Ray, R. L., Sargani, G. R., Ihtisham, M., Khayyam, M., & Ismail, S. (2021). Current progress and future prospects of agriculture technology: Gateway to sustainable agriculture. *Sustainability*, *13*(9), 4883. https://doi.org/10.3390/su13094883.

Khuman, S. N., Park, M. K., Kim, H. J., Hwang, S. M., Lee, C. H., & Choi, S. D. (2022). Organochlorine pesticides in the urban, suburban, agricultural, and industrial soil in South Korea after three decades of ban: Spatial distribution, sources, time trend, and implicated risks. *Environmental Pollution*, *311*, 119938. https://doi.org/10.1016/j.envpol.2022.119938.

Kim, K. H., Kabir, E., & Jahan, S. A. (2017). Exposure to pesticides and the associated human health effects. *Science of the Total Environment*, *575*, 525–535. https://doi.org/10.1016/j.scitotenv.2016.09.009.

Kumar, K., Gambhir, G., Dass, A., Tripathi, A. K., Singh, A., Jha, A. K., Yadava, P., Choudhary, M., & Rakshit, S. (2020). Genetically modified crops: Current status and future prospects. *Planta*, *251*(4), 91. https://doi.org/10.1007/s00425-020-03372-8.

Kunstadter, P. (2007). *Pesticides in Southeast Asia: Environmental, Biomedical, and Economic Uses and Effects*. Silkworm Books, Ms Trasvin.

Lekei, E., Ngowi, A. V., Kapeleka, J., & London, L. (2020). Acute pesticide poisoning amongst adolescent girls and women in northern Tanzania. *BMC Public Health*, *20*, 303. https://doi.org/10.1186/s12889-020-8374-9.

Li, Y. F. (1999). Global technical hexachlorocyclohexane usage and its contamination consequences in the environment: from 1948 to 1997. *Science of the Total Environment*, *232*(3), 121–158. https://doi.org/10.1016/S0048-9697(99)00114-X.

Liu, C. J., Men, W. J., Liu, Y. J., & Zhang, H. (2002). The pollution of pesticides in soils and its bioremediation. *System Sciences and Comprehensive Studies in Agriculture*, *18*(4), 295–297.

Liu, Y., Pan, X., & Li, J. (2015). A 1961–2010 record of fertilizer use, pesticide application and cereal yields: a review. *Agronomy for Sustainable Development*, *35*, 83–93. https://doi.org/10.1007/s13593-014-0259-9.

Lu, C., Hung, Y. T., & Cheng, Q. (2020). A review of sub-lethal neonicotinoid insecticides exposure and effects on pollinators. *Current Pollution Reports*, *6*, 137–151. https://doi.org/10.1007/s40726-020-00142-8.

Lu, C., Warchol, K. M., & Callahan, R. A. (2014). Sub-lethal exposure to neonicotinoids impaired honey bees winterization before proceeding to colony collapse disorder. *Bulletin of Insectology*, *67*(1), 125–130.

Main, A. R. (1979). Mode of action of anticholinesterases. *Pharmacology & Therapeutics*, *6*(3), 579–628. https://doi.org/10.1016/0163-7258(79)90066-4.

Marth, E., Stünzner, D., Köck, M., & Möse, J. R. (1989). Toxicokinetics of chlorinated hydrocarbons. *Journal of Hygiene, Epidemiology, Microbiology, and Immunology*, *33*(4 Suppl), 514–520.

Metcalf, R. L. (1955). *Organic insecticides. Their chemistry and mode of action*. Interscience Publ., Inc.

Narahashi, T., Frey, J. M., Ginsburg, K. S., & Roy, M. L. (1992). Sodium and GABA-activated channels as the targets of pyrethroids and cyclodienes. *Toxicology Letters*, *64*, 429–436. https://doi.org/10.1016/0378-4274(92)90216-7.

Narahashi, T., Ginsburg, K. S., Nagata, K., Song, J. H., & Tatebayashi, H. (1998). Ion channels as targets for insecticides. *Neurotoxicology*, *19*(4-5), 581–590.

Norton, L. B. (1948). Inorganic Insecticides. *Industrial & Engineering Chemistry*, *40*(4), 691–693. https://doi.org/10.1021/ie50460a023.

Oberemok, V. V., Laikova, K. V., Gninenko, Y. I., Zaitsev, A. S., Nyadar, P. M., & Adeyemi, T. A. (2015). A short history of insecticides. *Journal of Plant Protection Research*, *55*(3), 221–226.

Prasad, T. V. (2019). *Handbook of Entomology* (4th ed.). New Vishal Publications.

Rao, D. M., Anitha, S., Aravinda, A., Karunakar, B., & Devanna, N. (2012). Biopesticidal properties of seed, seed cake and oil of *Jatropha curcas* L. against the polyphagous lepidopteran pest *Helicoverpa armigera*. In N. Carels, M. Sujatha, & B. Bahadur (Eds.), *Jatropha, Challenges for a New Energy Crop: Volume 1: Farming, Economics and Biofuel* (pp. 369–382). Springer New York. https://doi.org/10.1007/978-1-4614-4806-8_20.

Reddy, D. S. (2018). *Applied Entomology* (3rd ed.). New Vishal Publications.

Rehman, H., Aziz, A. T., Saggu, S., Abbas, Z. K., Mohan, A., & Ansari, A. A. (2014). Systematic review on pyrethroid toxicity with special reference to deltamethrin. *Journal of Entomology and Zoology Studies*, *2*(6), 60–70.

Ren, Z., Cai, T., Wan, Y., Zeng, Q., Li, C., Zhang, J., Ma, K., He, S., Li, J., & Wan, H. (2023). Unintended consequences: Disrupting microbial communities of *Nilaparvata lugens* with non-target pesticides. *Pesticide Biochemistry and Physiology, 194*, 105522. https://doi.org/10.1016/j.pestbp.2023.105522.

ReportLinker. (2023). *Pesticide Market Outlook 2022 – 2026*. https://www.reportlinker.com/clp/global/13.

Sarkar, S., Narang, A., & Singh, S. (1994). Transplacentally acquired carbamate insecticide (Baygon) poisoning in a neonate. *Indian Pediatrics*, *31*(3), 343–346.

Shahid, M., & Khan, M. S. (2022). Ecotoxicological implications of residual pesticides to beneficial soil bacteria: A review. *Pesticide Biochemistry and Physiology, 188*, 105272. https://doi.org/10.1016/j.pestbp.2022.105272.

Sharma, A., Kumar, V., Shahzad, B., Tanveer, M., Sidhu, G. P. S., Handa, N., Kohli, S. K., Yadav, P., Bali, A. S., Parihar, R. D., Dar, O. I., Singh, K., Jasrotia, S., Bakshi, P., Ramakrishnan, M., Kumar, S., Bharadwaj, R., & Thukral, A. K. (2019). Worldwide pesticide usage and its impacts on ecosystem. *SN Applied Sciences, 1,* 1446. https://doi.org/10.1007/s42452-019-1485-1.

Shattuck, A., Werner, M., Mempel, F., Dunivin, Z., & Galt, R. (2023). Global pesticide use and trade database (GloPUT): New estimates show pesticide use trends in low-income countries substantially underestimated. *Global Environmental Change*, *81*, 102693. https://doi.org/10.1016/j.gloenvcha.2023.102693.

Shepard, H. E. (1939). *The chemistry and toxicology of insecticides*. Burgess Pub. Co.

Simon-Delso, N., Amaral-Rogers, V., Belzunces, L. P., Bonmatin, J. M., Chagnon, M., Downs, C., Furlan, L., Gibbons, D. W., Giorio, C., Girolami, V., Goulson, D., Kreutzweiser, D. P., Krupke, C. H., Liess, M., Long, E., McField, M., Mineau, P., Mitchell, E. A. D., Morrissey, C. A., Noome, D. A., … Wiemers, M. (2015). Systemic insecticides (neonicotinoids and fipronil): Trends, uses, mode of action and metabolites. *Environmental Science and Pollution Research*, *22*, 5–34. https://doi.org/10.1007/s11356-014-3470-y.

Smith, A. G. (2012). DDT and other chlorinated insecticides. In T. Marrs (Ed.), *Mammalian Toxicology of Insecticide* (pp. 37–103). RSC publishing, Cambridge. https://doi.org/10.1039/9781849733007-00037.

Statista. (2024, 18 October). Leading pesticide companies worldwide in 2023 based on sales. https://www.statista.com/statistics/257489/ranking-of-leading-agrochemical-companies-worldwide-by-revenue/.

Straisresearch. (2023). *Insecticides market*. https://straitsresearch.com/report/insecticides-market

Sugimoto, H., Iimura, Y., & Yamanishi, Y. (2002). The New Generation of Acetylcholinesterase Inhibitors. In Y. Mizuno, A. Fisher, & I. Hanin (Eds.), *Mapping the Progress of Alzheimer's and Parkinson's Disease* (pp. 193–198). Springer, Boston, MA. https://doi.org/10.1007/978-0-306-47593-1_32.

Tang, W., Wang, D. I., Wang, J., Wu, Z., Li, L., Huang, M., Xu, S., & Yan, D. (2018). Pyrethroid pesticide residues in the global environment: An overview. *Chemosphere*, *191*, 990–1007. https://doi.org/10.1016/j.chemosphere.2017.10.115.

Thatheyus, A. J., & Selvam, A. G. (2013). Synthetic pyrethroids: toxicity and biodegradation. *Applied Ecology and Environmental Sciences*, *1*(3), 33–36. https://doi.org/10.12691/aees-1-3-2.

Tlak Gajger, I., & Dar, S. A. (2021). Plant allelochemicals as sources of insecticides. *Insects*, *12*(3), 189. https://doi.org/10.3390/insects12030189.

Tomizawa, M., & Casida, J. E. (2003). Selective toxicity of neonicotinoids attributable to specificity of insect and mammalian nicotinic receptors. *Annual Review of Entomology*, *48*(1), 339–364. .https://doi.org/10.1146/annurev.ento.48.091801.112731.

Tomizawa, M., & Casida, J. E. (2005). Neonicotinoid insecticide toxicology: mechanisms of selective action. *Annual Review of Pharmacology and Toxicology*, *45*(1), 247–268. https://doi.org/10.1146/annurev.pharmtox.45.120403.095930.

Tsai, W. -T. (2024). Organochlorine insecticides. In P. Wexler (Ed.), Encyclopedia of Toxicology (4th ed., 7th vol., pp. 167–171). Academic Press. https://doi.org/10.1016/B978-0-12-824315-2.00970-2.

United Nations. (2015). *Resolution adopted by the General Assembly on 25 September 2015. 70/1 transforming our world: The 2030 agenda for sustainable development* (Report No. A/RES/70/1).

United Nations. (2022). *World population prospects 2022* (UN DESA/POP/2021/TR/NO.3).

Vale, A., & Lotti, M. (2015). Organophosphorus and carbamate insecticide poisoning. In M. Lotti & M. L. Bleecker (Eds.), *Handbook of Clinical Neurology* (131st vol., pp. 149–168). Elsevier. https://doi.org/10.1016/B978-0-444-62627-1.00010-X.

Worldometer. (2023, May 11). *Pesticide use by country*. Food and Agriculture, Pesticides, Woldometer. Retrieved May 11, 2023, https://www.worldometers.info/food-agriculture/pesticides-by-country/.

Zhang, X., Huang, Y., Chen, W. J., Wu, S., Lei, Q., Zhou, Z., Zhang, W., Mishra, S., Bhatt, P., & Chen, S. (2023). Environmental occurrence, toxicity concerns, and biodegradation of neonicotinoid insecticides. *Environmental Research*, *218*, 114953. https://doi.org/10.1016/j.envres.2022.114953.

Chapter 2

Overview of Insecticide Detection Techniques

Abran Singh Kushwah[1,*]
Anup Kumar[2]
Sonu Kumar Mahawer[3]
Neeraj Patanjali[1]
and S. Rajna[4]

[1]Division of Agricultural Chemicals, ICAR-Indian Agricultural Research Institute, New Delhi, India
[2]Division of Plant Animal Relationship, Indian Grassland Fodder Research Institute, Jhansi, India
[3]Division of crop production, ICAR- Indian Grassland Fodder Research Institute, Jhansi, India
[4]Division of Entomology, ICAR-Indian Agricultural Research Institute, New Delhi, India

Abstract

Detection and quantification of insecticides have become very important due to their toxic nature towards biotic and abiotic ecosystems. This chapter discusses the principles, strengths, and limitations of various techniques for detecting insecticides using chromatographic, spectrometric, immunoassay, and electrochemical methods. These techniques are widely accepted globally in terms of toxicity and environmental safety. Paper chromatography, thin-layer chromatography (TLC), high-performance liquid chromatography (HPLC), and gas chromatography (GC) are some of the chromatographic techniques used for pesticide identification. Among spectroscopic techniques, Mass Spectrometry (MS), Infrared spectrometry, and UV-visible spectrometry

---

* Corresponding Author's Email: abransinghagrian@gmail.com

In: Insecticides
Editors: Archana Anokhe and Govindaraj Kamalam Dinesh
ISBN: 979-8-90134-022-6

are used for the detection and quantification of insecticides. However, with advancements in technology and the multi-residue analysis of insecticides, coupled with further lowering of the detection limit, a combined instrument of chromatography coupled with spectroscopic detection is now becoming very popular in insecticide detection. They are HPLC/UPLC and GC coupled with mass spectrometry (GC-MS/MS or LC-MS/MS), where HPLC/UPLC and GC are used for compound separation, and MS serves as the detector. These instruments are significantly more advanced in terms of time consumption, detection limit, and solvent consumption. Immunoassay may be used in agriculture to check the presence of insecticides in grain or animal feed. For many insecticides, ELISA tests were successfully developed. Another detection technique, named electrochemical detection, is now emerging, which can detect the pesticide at the site. This technique uses electricity for the movement and detection of target analytes. The detection limits are comparable to the above spectroscopic and chromatographic techniques.

Keywords: insecticide detection, residue, chromatography, spectroscopy, and immunoassay

Introduction

Insecticides help to increase the total output and production of agricultural foods by reducing the number of insect pests. However, excessive use of insecticides has led to the contamination of food, water, and the environment, as well as the disruption of biological systems. Pesticides are dangerous substances that can influence people over time, leading to significant health problems. To ensure the availability of healthy foods in the food supply system and to help farmers develop the most effective agronomic crop production methods, international health authorities have established limits for pesticide residues in specific food products. Traditional detection methods, such as chromatography and spectroscopy, are highly accurate and deliver precise results with excellent sensitivity and selectivity; however, they require more time and cost. The commercial viability of electrochemical and immunoassay approaches is relatively high. Different sensors can be used to detect many analytes simultaneously (Saha et al., 2023; Sindhu and Manickavasagan, 2023). To enhance sensitivity, accuracy, and recovery efficiency prior to detecting pesticide residues, sample pretreatment is crucial. Sample preparation typically entails extraction, techniques for pre-concentration, separation, and others. The immunoassay approach has been utilized extensively in the detection of pesticide residues in samples compared to older detection techniques due to its high sensitivity, quick turnaround time, straightforward operation, and high specificity. The enzyme-linked immunosorbent assay (ELISA) is now one of the widely used immunoassay methods for detecting pesticide residues. The results of ELISA, GC, and GC-MS were well correlated, indicating that ELISA is a reliable substitute for the expensive and time-consuming GC approach (Miao et al., 2023).

Chromatography

Chromatography is a well-known laboratory technique used for the separation and measurement of complex chemical mixtures that other purification procedures cannot efficiently separate. The components of a mixture dissolved in a solvent get gradually removed from one another by the mobile phase in a gradient manner according to their affinities to the stationary phase. Mikhail Semenovich Tswett developed chromatography in 1903 while studying plant pigments like chlorophyll, xanthophyll, and carotenoids (Ettre, 2003). These methods were further improved by research conducted in the 1940s and 1950s by Archer John Porter Martin and Richard Laurence Millington Synge, for which they were awarded the 1952 Nobel Prize in Chemistry. Nowadays, there are many distinct types of chromatography techniques, such as thin-layer chromatography (TLC), paper chromatography, liquid chromatography (e.g., HPLC, UPLC, and preparative HPLC), supercritical fluid chromatography, and gas chromatography, which have been developed and used for the separation and purification of compound mixtures (Williams, 1979). Liquid chromatography (including HPLC, UPLC, and LC-MS) is a frequently used characterization technique in the detection of insecticides. HPLC and UPLC can be used as quantitative approaches to clarify the structure of the molecule and facilitate quantification when connected with a mass detector (MS). Each one differs in the stationary phase and operational parameters.

Chromatography can be run in two different ways: regular and reverse phase. Depending on the kind of sample that needs to be separated, the mode selection is crucial. Most molecules are best separated using reversed-phase chromatography, which uses polar mobile phases and nonpolar stationary phases; details of the mobile phase and stationary phase are given in Table 1. Isomer (enantiomer) separation, however, requires the use of normal-phase chromatography, which employs nonpolar mobile phases and polar stationary phases. Based on their hydrophobicity, reversed-phase chromatography separates the components with good resolution. The more polar the molecule, the faster it elutes, and the less polar the compound, the slower it elutes. The general classification of chromatography is described in Figure 1.

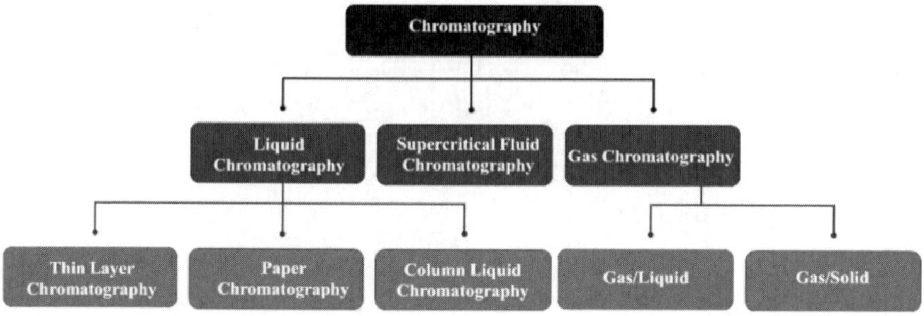

Figure 1. A plan for categorizing chromatography into different applicable approaches.

Liquid Chromatography (LC)

Principle: Liquid chromatography (LC), frequently referred to as high-performance LC (HPLC), is a technique used in many analytical applications. High pressure, defined as 2000–6000 psi, was another term for it in the past. They deal with the variance in interactions between the analytes, or target species, and the mobile and stationary phases. A successful separation can be accomplished by carefully selecting both of its components. The following formula can be used to express the analyte's equilibrium distribution coefficient (K) between the two phases.

$$k = Cs/Cm$$

where *Cs is the stationary phase* and *Cm* is the mobile phase.

Types of Liquid Chromatography (LC):

1. Partition (liquid-liquid) chromatography: Liquid-liquid chromatography, commonly known as partition chromatography, was developed in 1940 by Martin and Synge. The distribution coefficients of the analytes between the stationary phase and the mobile phase determine partition chromatography. If there is a significant

difference between the distribution coefficients, separation occurs. (Hamilton and Sewell, 1982).

2. Normal-phase chromatography: In normal-phase chromatography, the stationary phase is polar, and the mobile phase is comparatively nonpolar (Wang, 2007).
3. Reversed phase chromatography: Reversed phase chromatography is one of the most widely used types of LC, also known as HPLC (RPLC). The stationary phase must be comparatively nonpolar than the mobile phase for the separation to occur (Michalke, 2002).
4. Ion exchange chromatography (IEC): Ion exchange chromatography has proven to be a widely used technique since its inception. Ionic groups are present in column packing, and the buffer serves as the mobile phase. It functions by using a charged stationary phase that attracts the opposing charge and exchanges it with a counter-ion. Anion exchange involves a stationary phase (R+) with a net positive charge. The subsequent equilibrium is reached (Hamilton and Sewell, 1982).

$$R^+Y^- + X^- = R^+X^- + Y^-$$

where Y^- is the attracted negative counter-ion on the stationary phase, and X^- is a sample anion in the mobile phase.

5. Size exclusion chromatography: The earliest commercial application of HPLC was gel permeation chromatography, commonly referred to as size exclusion chromatography (SEC). Because its separation is dependent on molecule size, it is regarded as the most basic form of chromatography. Depending on the column packing, if the compound is too massive to fit into any of the stationary phase pore spaces, it will be washed through the void space, sometimes referred to as the exclusion volume. Because they can fit through pores, smaller molecules have a longer path to travel and elute later than larger molecules. The inclusion volume is the point at which all species elute concurrently at the end because the compounds' size and diameter are so small. There may be molecules of different sizes present in a single system (McMaster, 2007).

Thin Layer Chromatography (TLC): It is based on the principles of adsorption chromatography, partition chromatography, or a combination of the two, depending on the adsorbent, its handling, and the type of solvents used. Components that are more attracted to stationary phases move more slowly. Moving parts move more quickly if they have a lower attraction for the stationary phase. When normal-phase solvents offer the best separation, using TLC is an excellent option. The usual glass plates used for thin-layer separations have a thin coating of stationary phase on them. Adsorption, normal- and reverse-phase, ion exchange, and size exclusion are only a few of the chromatographic techniques covered by the stationary phases used in TLC. Due to its simplicity, TLC is an ideal method for isolating chemicals. The impurity and degradant level must be at or above 1%, though, for TLC to be effective. Due to the greater detection limits, any component present at this level or below is particularly challenging to isolate on a TLC plate (Ahuja and Scypinski, 2001).

These are the steps involved in TLC developments (Figure 2):

1. Applying the samples to the plates
2. Developing the plates
3. Identifying and locating the target compound
4. Extracting the target compound

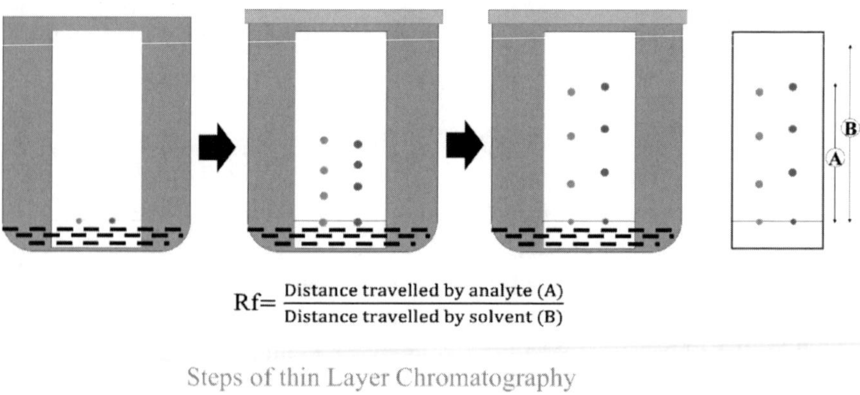

Figure 2. Pictorial representation of Thin Layer Chromatography.

Retention factor (R_f) value: Individual compounds emerge as vertically separated spots following a successful separation. The migration distance of the compound divided by the solvent's total distance travelled reflects the retention factor of each location. The retention factor is always less than 1. The less polar compound will get a greater value of R_f because it does not adhere to the stationary phase for as long as the polar substance, which would have a lower R_f value when comparing two different substances under the same circumstances.

$$R_f = \frac{\textit{Distance traveled by the compound } (a)}{\textit{Distance traveled by the solvent } (b)}$$

UV light is frequently used for detection. When the desired separated compound has been located on the plate, the band is scraped, and the impurity is removed using the proper solvent from the stationary phase (silica gel). The extracted substance is filtered or centrifuged, and the collected solvent is evaporated to produce the isolated material. It does not need expensive or complex analytical equipment. The simultaneous analysis of many samples opens up the potential for a wide range of detection techniques for the separated components, including UV or chemical reactions that produce colored spots (Valko, 2000).

Paper Chromatography: In 1944, paper chromatography was developed. Both the stationary phase and the mobile phase used in partition chromatography are liquids. The liquid stationary phase is typically supported by paper. The dissolved sample is placed as a tiny spot or streak at least half an inch from the edge of a strip of filter paper, often made of cellulose, and left to dry.

Table 1. Commonly occurring mobile and stationary phases (Cornelis et al., 2003; Wang, 2007)

Type of LC	Mobile phase	Stationary phase
Normal phase	Nonpolar mobile phase: hexane, chloroform, diethyl ether, etc.	Polar packing: usually silica
Reverse phase	Water with organic modifier (i.e., methanol or acetonitrile)	Nonpolar stationary phase (C18 or C8)
Ion exchange	Polar mobile liquid with water, salts, or alcohol	Fixed acidic or basic sites ($-SO_3^-$, $-NR_3^+$) on beads of cross-linked styrene and divinylbenzene
Size exclusion	Selected based on the ability to solubilize the analyte being separated	Aerogels, xerogels, and aerogel-xerogels

High-performance liquid chromatography (HPLC): The instrumentation of HPLC is described with a block diagram in Figure 3.

Instrumentation of HPLC:

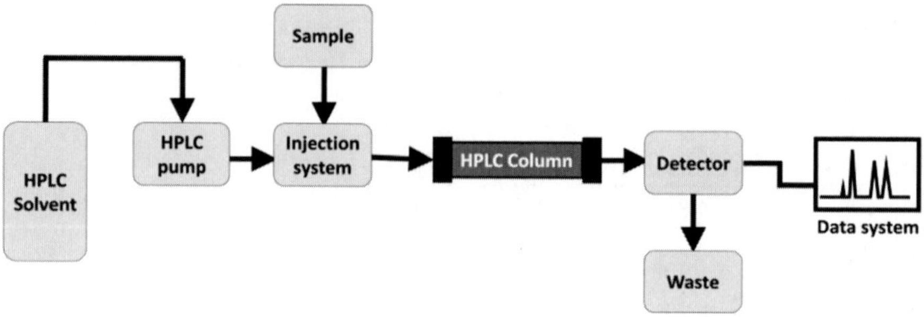

Figure 3. Block diagram of HPLC system.

Pump: There are often two different modes that can be used, depending on the mobile phases. Suppose the composition of the mobile phase remains constant during the separation process. In that case, the HPLC system is referred to as an isocratic elution system. When the composition of the mobile phase changes during separation, the HPLC system is referred to as a gradient elution system. A gradient system has two ways: a low-pressure gradient (LPG) and a high-pressure gradient (HPG). If the gradient is low, the solvents are mixed on the suction side of the pump, upstream from the motor. In a high-pressure gradient system, various solvents are delivered by separate pumps and combined on the discharge side after the pumps have been used.

Injection valve: The HPLC system receives the sample itself through the injection valve. The injection valve also uses high pressure because it is situated between the pump and the columns. The figure depicts a 6-port one in a schematic. A sample aliquot (~20 μL) is injected into the sample loop during the loading stage of the injection valve, and the additional sample is used to rinse the loop before it exits to the waste line. The mobile phase moves the sample through the columns in the injection position. The loop can be manually filled with the sample using a syringe; however, it should be washed in between subsequent injections to ensure that nothing is left over that could contaminate the next injection. However, an autosampler is often employed for the analysis of a large number of samples. Additionally, it offers the benefit of enabling nightly autonomous analyses (McMaster, 2007). For a better understanding of the injection valve, please refer to Figure 4.

Column: The column is the beating heart of any HPLC machine. It is responsible for correctly partitioning the sample components. The sort and size of packing material used in the column, its inner diameter, and its length all have an impact on the effectiveness of separation. Depending on the intended usage, many commercial HPLC columns are available. Different packing materials support diverse separation strategies.

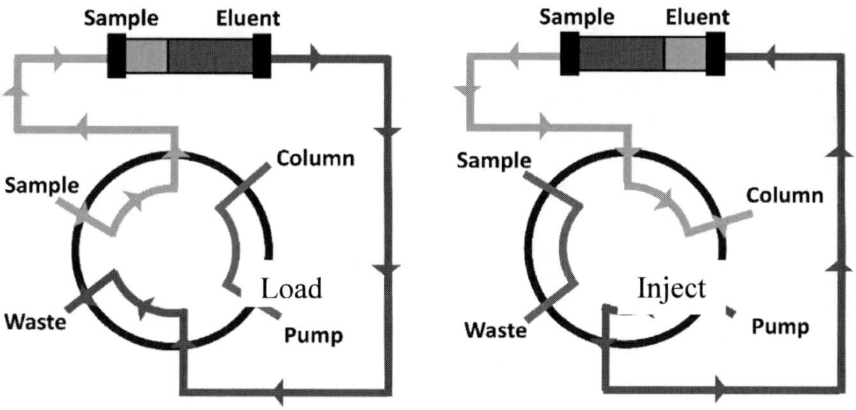

Figure 4. Injection system of HPLC.

Detector: The last HPLC component, the detector separates bands or analytes that have been eluted from the column via the mobile phase. HPLC has a variety of detectors, which have been listed in Table 2.

Table 2. List of HPLC detectors with target compounds

Detector	Target Compounds
Absorbance(Ultraviolet Detectors & Photo Diode Array)	UV-absorbing compounds
Fluorescence	Fluorescent compounds
Refractive Index	All
Evaporative Light Scattering	Al, exclude non-volatile compounds
Conductivity	Cation, Anion
Mass Spectrometry	Ionic compounds

Chromatogram: The signal peaks of the separated analytes, carried by the mobile phase, are recorded by the detector unit. The total of all the peaks is referred to as the "chromatogram" (Figure 5). Both qualitative and quantitative information about the analyte can be found in each unique peak. Qualitative information (such as shape, signal intensity, and chromatogram appearance time) is provided by the peak itself. The concentration of the drug is also correlated with the area of a peak. Thus, through integration, the chromatography data management software can ascertain the sample's concentration. This displays numerical information. The ideal representation of the peaks will be a Gaussian bell-shaped curve.

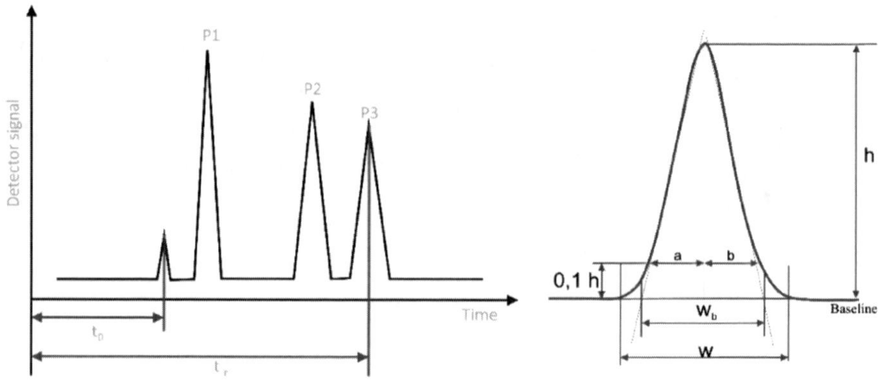

Figure 5. Schematic illustration of a chromatogram.

Time of delay (t_0): The time it takes for a non-retarded chemical to travel from the injection site to the detection device, where it is recorded, is known as the delay time. At this moment, every sample molecule is only present in the mobile phase. Generally, the delay duration of each sample molecule is the same.

Time for Retention (t_R): In High-Performance Liquid Chromatography (HPLC), retention time refers to the time it takes for a particular compound to travel through the chromatographic column from the point of injection to the point of detection. HPLC is a widely used analytical technique in which a liquid mobile phase is pumped through a packed column containing the stationary phase.

Peak width (w): The period from the start of the signal slope to reaching the baseline following repeated drops in the detector signal is covered by the peak width.

Tailing factor (T): Peaks frequently display some tailing in a chromatogram. The tailing factor T is used to calculate peak tailing. This factor indicates the degree to which the shape resembles a fully symmetric Gaussian curve at its peak. The measurements should be at 10% of the peak height.

$$T = b/a$$

where a is the breadth of the front half of the peak, b indicates the width of the back half.

$T = 1$: symmetrical peak, $T > 1$: tailing, $T < 1$: fronting.

To analyze insecticide residues, a variety of liquid chromatography-based methods have been proposed, with the majority incorporating UV, photodiode array (PDA), diode array detector (DAD), and mass spectrometry (MS) detectors. Even though liquid chromatography, when combined with UV, PDA, and DAD technologies, is employed, it remains challenging to provide structural information on chemical compounds for identification. Due to this, structural interventions have been overcome by utilizing the mass detection technique, which also employs tandem mass spectrometry (MS/MS) to obtain structural information from molecular masses and fragmentation patterns. Utilizing various columns of reverse phase (Carbon-8, Carbon-12, and Carbon-18), organic phases, and buffers (acids of formic acid, ammonium formate, acetic acid, and ammonium acetate), experiments have been conducted using liquid chromatography with mass detection. Sometimes, gradient mode is used, with flow rates generally ranging from 0.2 to 1.0 mL/min, depending on the analysis. Solvent mixes (water-acetonitrile and water-methanol) are used in these situations (Christia et al., 2015); A (Jallow et al., 2017). Octadecyl C18 is the most commonly used stationary phase for chromatographic separation, and gradient mode (dynamic solvent mixture as the mobile phase) has been employed for multi-residue analysis to shorten the runtime. Methanol and acetonitrile are the two solvents used in reverse-phase chromatography the most frequently. However, tetrahydrofuran (THF) is utilized less frequently (Joshi et al., 2015).

Ionization sources for mass detection, such as electrospray ionization (ESI), are frequently employed because they can efficiently charge both polar and nonpolar analytes (Dzuman et al., 2015). For both qualitative and quantitative examinations, the mass analyzers Q-Trap (Crnogorac et al., 2008) and triple quadrupole (QqQ) (Rong et al., 2018) are suitable. Apart from these analyzers, a single quadrupole can also be used for multi-residue analysis (Bakirci and Hişil, 2012).

Advantages of HPLC:

- Speedy
- Efficiency is very high

- Excellent accuracy is achieved through the overall broad range of detection techniques, including sensitivity, selectivity, repeatability, and versatility, which make HPLC an effective instrument for chemical investigation across a multitude of domains. Limitations of HPLC:
- HPLC is expensive as it involves significant amounts of pricey organic substances.
- Complexity
- Certain chemicals are only slightly detectable, while others are permanently adsorbed, making them challenging to identify using HPLC.
- GC is superior at separating volatile compounds.

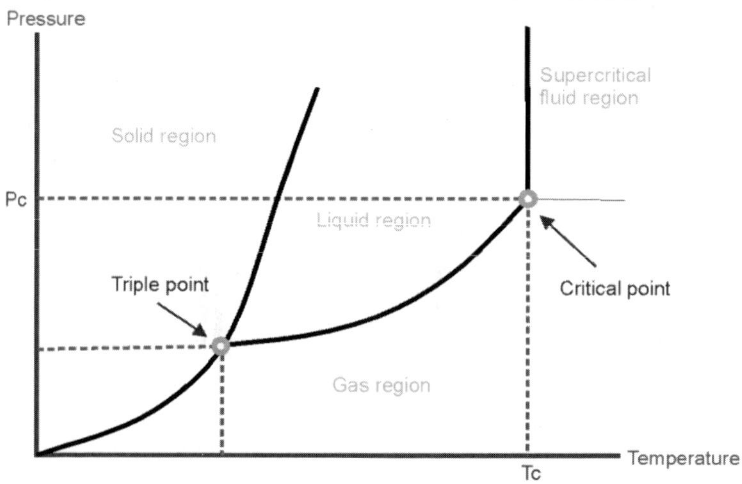

Figure 6. Illustration of the supercritical fluid region.

Supercritical Fluid Chromatography (SFC): SFC refers to chromatography performed above the critical pressure (Pc) and critical temperature (Tc) of the mobile phase. Neither a liquid nor a regular gas is a supercritical fluid. The critical point is the intersection of Pc and Tc (Figure 6). By increasing the pressure or by boosting the temperature, a conventional liquid or gas can be converted into a supercritical fluid. Although carbon dioxide is widely employed as the mobile phase for SFC, polar and high-molecular-weight molecules do not dissolve well in it. A nonpolar supercritical fluid can benefit from a little addition of a polar, organic solvent like methanol to improve peak form, increase solute solubility, and change selectivity (Jumhawan and Bamba, 2010).

Supercritical fluids give chromatographic properties that are between LC and GC. In comparison to LC, supercritical fluids' high diffusivity and low viscosity result in shorter analysis periods and better resolution.

Gas chromatography (GC):

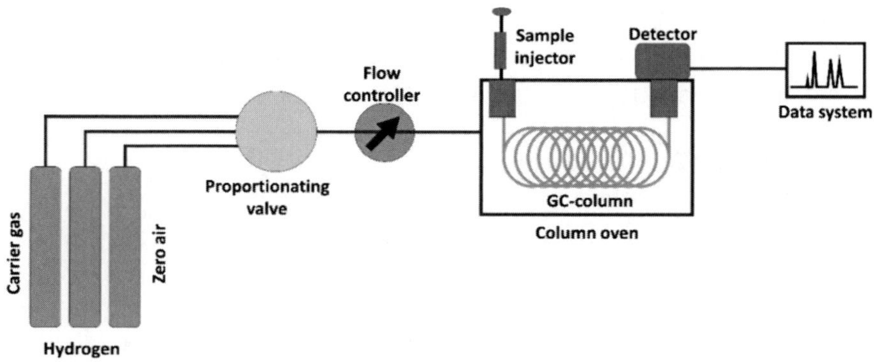

Figure 7. Pictorial diagram of Gas chromatography.

Carrier gas: The carrier gas must be chemically inert. Gases, including carbon dioxide (CO_2), nitrogen (N_2), helium (He), and argon (Ar), are commonly used in this process. The type of detector often influences the choice of carrier gas. The carrier gas system additionally includes a molecular sieve to filter out water and other contaminants. Pictorial representation of the gas chromatography is mentioned in the Figure 7.

Sample injection port: In comparison to syringe injection, gas sampling valves offer injection precision that is at least ten times higher. Multiport valves are frequently utilized. The primary benefit of gas switching valves is the elimination of operator mistakes in sample volume intake, which can occur with syringe injections. Additionally, while the sample is being delivered into the empty sample loop, the carrier gas flow is not halted. The sample loop's contents are injected into the carrier gas stream when the valve is switched. Using injector valves, which can be controlled manually or automatically, sample volumes can be introduced in a reproducible way (Jennings and Mehran, 1986).

Columns: The components of a sample are separated when it flows through a column of inert gas used as the mobile phase, due to variations in how they interact with the stationary phase (Rahman et al., 2015).

Packed GC columns: A slight coating of liquid is adsorbed either onto the solid inert packing or the surface of the beads within the column. Applicable to both GSC and GLC. Most silicone-based oils used in GC liquid phases have excellent thermal stability. Methyl silicone, methyl phenyl silicone, methyl trifluoropropyl silicone, and methyl cyanoethyl silicone in a variety of polarities.

Capillary GC Columns: The inner wall of capillary chromatographic columns is not packed with material; instead, a thin layer of liquid phase is applied. Due to the tube's open design and minimal flow resistance, it is classified as an open-tubular column.

a. Porous Layer Open Tubular Column
b. Wall-Coated Open Tubular Column
c. Support-Coated Open Tubular Column
d. Fused Silica Open Tubular Column

Detector: The signal from a concentration-dependent detector is correlated with the solute concentration in the detector and typically does not cause the sample to be destroyed. The responsiveness of the detectors will be reduced by dilution with makeup gas. Mass flow-dependent detectors typically destroy the sample, and the signal is correlated with the rate at which solute molecules enter the detector. Makeup gas has no impact on the response of a mass flow-dependent detector. Details of the GC detector have been described in Table 3.

The sensitivity of specific detectors, such as the electron-capturing detector (ECD), flame photometric detector (FPD), nitrogen phosphorus detector (NPD), and mass selective detector (MSD), makes them suitable for usage in combination with GC (Łozowicka et al., 2017) (Bakirci and Hişil, 2012). TCD, FID, and MS are universal detectors. Although MS typically operates in high vacuum, element-specific detectors like the ECD require temperatures above 300°C for optimum performance and auxiliary gas flow of nitrogen or argon/methane. To increase method sensitivity, mass detection techniques are also employed. These techniques utilize analyzers such as the ion trap (IT), quadrupole, triple quadrupole (QqQ), and time-of-flight mass analyzer (TOF) (Cavalcante et al., 2010; Reichenbach et al., 2015).

Table 3. Details of gas chromatography detectors

Detector		Type	Support gases	Selectivity	Detection limit
Universal	Flame Ionization Detector (FID)	Mass flow	H_2 and air	Organic compound	100 Pg
	Thermal Conductivity Detector (TCD)	Concentration	Reference	Universal	1 Ng
	Mass Spectrometer (MS)	-	-	Ionized molecule	10 Pg to 10 Ng
Selective	Electron capture (ECD)	Concentration	Makeup	Halides, nitrates, nitriles, peroxides, anhydrides, organometallics	50 Fg
	Nitrogen-phosphorus (NPD)/FTD	Mass flow	H_2 and air	N, P	10 Pg
	Flame photometric (FPD)	Mass flow	H_2 and air, possibly O_2	S, P, Sn, B, As, Ge, Se, Cr	100 Pg
	Photo-ionization (PID)	concentration	Makeup	Aliphatics, aromatics, ketones, esters, aldehydes, amines, heterocyclics, organosulphur, organometallics	2 Pg

The multiple Detectors can be connected to the GC in series or parallel; however, they are not frequently connected in series. Similar operating parameter requirements should be shared by two detectors connected in series, and the first detector in the series must not be destructive. A serial detector connection provides several benefits; that is, one GC detector port is used, and there is no transfer line or detector splitter (which can affect peak shape and performance). The sample is not divided; instead, the entire sample is analyzed. In the case of parallel detector conjunctions, which function independently of one another, it reduces difficulties driven by variations in the operating parameters required for the various detectors (Dąbrowski, 2018).

Spectrometry is a technique used to analyze the properties of matter by measuring the interactions between electromagnetic radiation and matter. It involves the study of spectra, which are characteristic patterns of wavelengths emitted, absorbed, or scattered by substances.

EMR or Electromagnetic Radiation

Frequency, wavelength, wave number, and Energy are used to describe electromagnetic waves. When an electric field accelerates a subatomic particle, like an electron, EM radiation is produced. Magnetic and electrical fields have oscillated as a result of the movement, and they move perpendicularly to one another and the direction of the wave (Figure 8). A detailed study of electromagnetic waves can be conducted effectively by correlating Figures 8, 9, and Table 4.

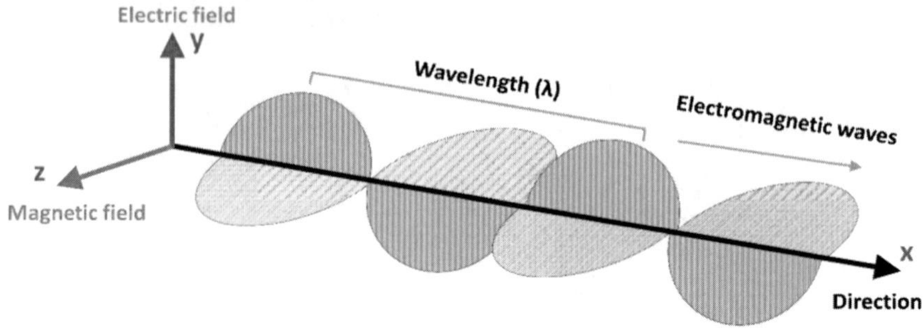

Figure 8. Electromagnetic waves.

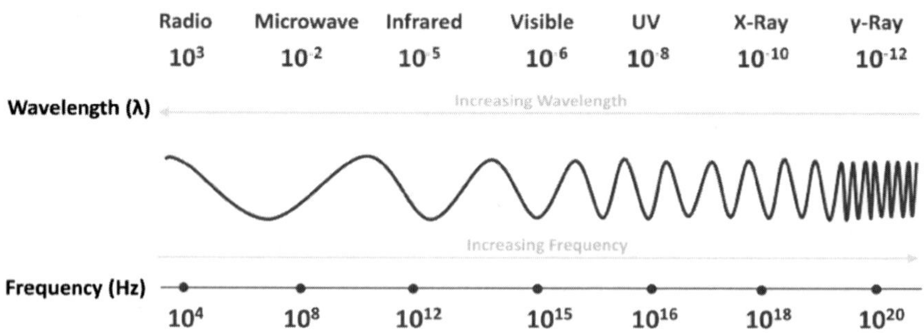

Figure 9. Electromagnetic waves with wavelength and Frequency.

Wavelength (λ) is the distance between two successive wave crests. Unit = m.

Frequency (ν) Frequency, in the context of spectrometry and electromagnetic radiation, refers to the number of cycles of a wave that occur in a given unit of time. Unit = hertz (Hz), or the number of wave cycles per second.

$$\nu = \frac{c}{\lambda}$$

c: light speed 3×10^{10} cm/sec; λ: wavelength cm.

Wavenumber ($\bar{\upsilon}$). It is the reciprocal of wavelength and is expressed in units of reciprocal length, such as reciprocal centimetres. Unit: cm^{-1}

The relation between wavelength, Frequency, and wavenumber is:

$$\frac{1}{\lambda} = \bar{\upsilon} = \frac{\nu}{c}$$

Table 4. Details of electromagnetic waves (Zwinkels, 2020)

Adsorption-induced transition		Wavelength range (nm)	Frequency range (s^{-1})	Description
Nuclear excitation	-	<0.1 nm	10^{20}–10^{23}	Gamma rays
Electronic transition	Inner electron (K & L shell)	0.1–10 nm	10^{17}–10^{20}	X-rays
	Outer electron (middle & valence shell)	10–400 nm	10^{15}–10^{17}	Ultraviolet
		400–700 nm	10^{14}–10^{15}	Visible
Change of configuration	Molecular vibrations	700 nm to 1 mm	10^{11}–10^{14}	Infrared
Change of orientations	Molecular rotations	1 mm to 1 cm	10^{10}–10^{11}	Microwaves
Change of spin (NMR)	Spin orientations	1 cm to 100 km	10^{3}–10^{10}	Radio waves
-	-	100–1,000 km	10^{2}–10^{3}	Audio frequency

The scientific field of spectroscopy examines how electromagnetic radiation interacts with materials, such as atoms and molecules. Atomic and molecular spectra are two different types of spectra that result from the interaction of EMR with matter. When the Energy of an atom changes, electrons move from one energy level to another, creating the atomic spectrum. In addition to electron transitions, the molecular spectrum also includes electron transitions between vibrational and rotational Energy Levels. Therefore, the molecular spectrum is much more complex than the atomic spectrum. By examining frequency, Wavelength, Wave number, Energy, and molecular processes, molecular spectroscopy provides a clear picture of how diatomic and polyatomic molecules interact. It also provides the most useful information about the shape and size of molecules, including bond angles, bond lengths, bond strengths, and bond dissociation energies.

Absorption and Emission Spectroscopy

When electromagnetic radiations fall within a specific wavelength range and travels through the substance being studied, specific wavelengths of radiation are absorbed by the substance. Absorption spectroscopy is the suitable term used for this type of research. Excitable electrons have a shorter existence time to return to their stable state and emit Energy. The study of this is called emission spectroscopy. The pictorial explanation of absorption and emission in the energy is mentioned in the Figure 10.

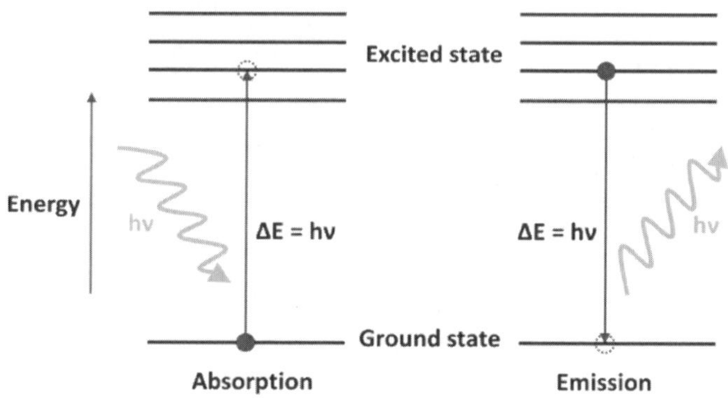

Figure 10. Absorption and emission in the energy diagram.

UV-Visible (UV-Vis) Spectrophotometry

Principle: The distance that light travels through the sample when it is exposed to light and the concentration of the target analyte that needs to be quantified are proportional to the absorbance. Beer's law is a formula that expresses this relationship (L. C. Passos and M. F. S. Saraiva, 2019):

$$A = \varepsilon \times b \times c$$

A: absorbance
ε: molar absorbance coefficient ($mol^{-1}\ cm^{-1}$)
b: route length (cm)
c: absorber concentration ($mol\ L^{-1}$).

The percentage of incident light that is transmitted by a sample is known as its transmittance (T):

$$T = \frac{I}{I_0}$$

where I, incident light, and I_O transmitted light.

The negative logarithm to base 10 of the transmittances is known as absorbance (A):

$$A = -log_{10}T = log_{10}\frac{1}{T} = log_{10}\frac{I_O}{I} = \varepsilon x a x b$$

Numerous compounds, including a large number of organic molecules, regularly absorb light in the UV-Vis range (~180 to 800 nm). Transitions n→σ* and σ → σ* require higher Energy and are associated with absorption in the UV region, while transition π →π* is associated with absorption in the UV-Vis area (Figure 11). Chromophore refers to the region of the molecule where the electronic transitions take place; UV or visible light is absorbed by a covalently unsaturated group, which may or may not affect the colour of the molecule. Ex. -C=C-,-N=N- (Reinecke, 2004). The groups that are saturated and unsaturated, and contain one or more pairs of unbonded electrons, are called auxochromes. Ex: -OH, $-NH_2$, -OR, *etc*. When this group is coupled to a chromophore, it helps to change the wavelength by intensifying the absorption and raising the λ_{max}.

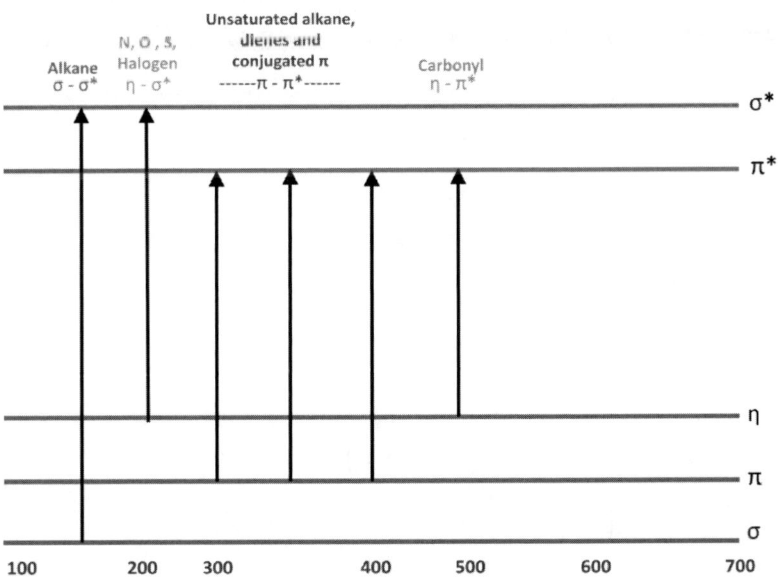

Figure 11. Different electronic transitions and their relation with wavelength (nm) and functional groups.

The absorbance of some materials may fluctuate over time for various reasons, such as photochemical reactions, aggregate formation, or adsorption on cell walls. Such alterations result in sample errors. Concentration, pH, time, temperature, presence of additional components, and solvent interactions are the key variables influencing absorption spectra and frequently result in variations from the Beer-Lambert law. These attenuations may occur as a shift in wavelength or as modifications to the position and shape of bands.

Bathochromic shift: The bathochromic shift, also known as a redshift, refers to a phenomenon in which the absorption or emission peak of a substance shifts towards longer wavelengths, resulting in a redder color. This shift occurs due to changes in the electronic structure or environment of the molecule.

Hypsochromic shift or Blueshift occurs when a compound's absorption maximum shifts to a shorter wavelength.

Hyperchromic shift: When the absorption band intensity increases.

Hypochromic shift: Decreased absorption band intensity.

Instrumentation of UV-VIS: Details of the different parts are provided below and are correlated with Figure 12.

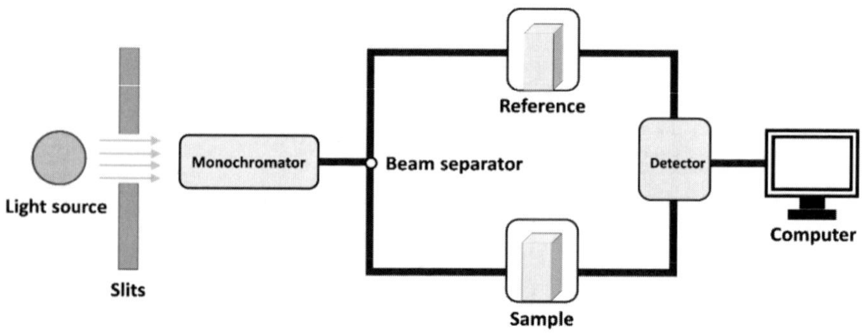

Figure 12. Pictorial representation of UV-Visual spectrometer.

Radiation source: The most popular source of radiation in the U.V. area (200–400 nm) is a hydrogen-discharge lamp, but deuterium-discharge lamps are utilized when higher intensities (3-5 times) are required. When determining absorption in the visible region (400-800 nm), a tungsten-filament lamp is utilized.

Monochromator: The radiation is divided into different wavelengths that only allow those at a particular wavelength to pass through it, which aids in their separation. The most common components of monochromators are quartz prisms or gratings. This is the case because quartz's ability to absorb radiation is limited, ensuring that the radiation retains its intensity and produces precise results.

Beam separator: Beam separators, as their name implies, assist in dividing a single beam of radiation into the reference chamber and the sample chamber. The 1st is known as the reference beam, while the 2nd is called the sample beam.

Detector: Photocells or photomultiplier tubes in detectors produce a voltage proportional to the Energy of the radiation that strikes them.

Recorder: The spectrum is automatically captured by a recorder as a plot of the wavelengths of the absorbed radiation against absorbance or molar absorptivity.

Advantages and applications of ultraviolet-visible spectroscopy:

1. Functional group detection.
2. Measuring the degree of conjugation.
3. Identifying unidentified substances.
4. Using UV-VIS spectrometers is simple.
5. The high accuracy of the instrument is crucial for chemists.

Limitations of Uv-Visual spectroscopy:

1. This method can only be used to study materials that absorb light in the visible or ultraviolet regions of the electromagnetic spectrum.
2. A substance's concentration may not be determined using UV-Vis spectroscopy if it does not have a known molar extinction coefficient.
3. UV-Vis spectroscopy is responsive to variations in pressure and temperature.
4. Light scattering can affect the precision of UV-Vis spectroscopy.

IR or Infrared Spectrometry

Principle: Diatomic molecules having permanent dipole moments, also known as heteronuclear atomic molecules and polyatomic molecules, are responsible for the IR spectra. IR radiations, which have a wavelength range of 14000 to 20 cm-1, are absorbed by atoms and molecules, causing them to vibrate and change vibrational states. The rotational energy levels change in tandem with the transitions in the vibrational energy levels. Rotational-vibrational spectra are another name for infrared spectra. These produce the atoms' and molecules' distinctive absorption bands. Different types of IR are given in Figure 13 and infrared spectroscopy block diagram is mentioned in the Figure 14.

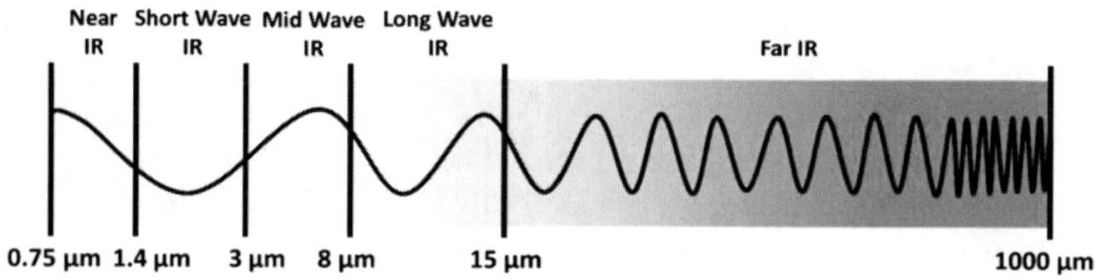

Figure 13. Different IR regions.

Instrumentation of IR spectroscopy:

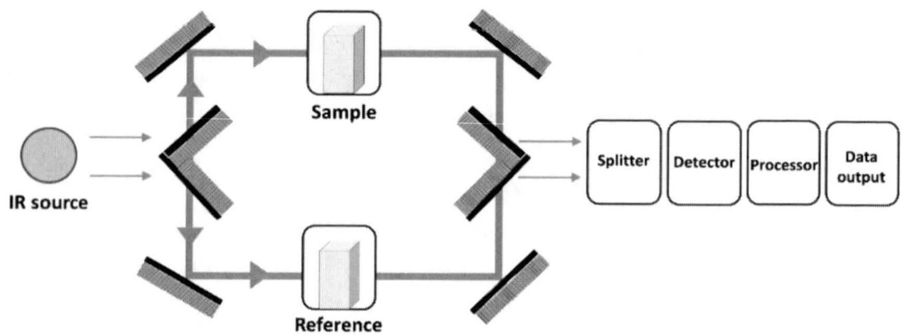

Figure 14. Block diagram of infrared spectroscopy.

Source of radiation: The source of radiation should produce Infrared radiation that is stretched throughout the desired wavelength, steady, and intense. For this reason, radiation is produced by sources such as globular sources, incandescent lamps, and mercury arc lamps, among others.

Monochromator: Since our sample will only absorb a limited range of IR frequencies, it is helpful to select the required IR frequencies. The most popular monochromators are the prism and grating.

Sample cells and sampling: The characterization of many materials, including solids, liquids, and gases, is performed using IR spectroscopy. Sampling of various phases using a substance that is transparent to IR light. Solids can be processed using pressed pellets, mull, and run-in processes.

Samples are placed in rectangular NaCl and KBr cells. That is liquid at room temperature. Similar to liquid sampling, gas sampling is done with a cell constructed of KBr or NaCl. Moisture must be prevented, and the gas should not react with the cell windows or the reflecting surfaces.

Detector: The detector aids in detecting the Infrared radiation that the sample transmits. The detector must be positioned and constructed to capture even low-frequency signals. Thermal detectors are, therefore, frequently utilized in IR spectroscopy. Detectors like the ones listed below are also used.

1. Bolometers 2. Photoconductive cell
3. Thermocouple 4. Semiconductivity cell
5. Thermistors 6. Pyroelectric detectors
7. Golay cell 8. Fourier Transform systems

Fundamental vibration: Stretching vibrations may be symmetrical, asymmetrical, and bending vibrations are in the plane and out of the plane, like scissoring, rocking, twisting, and wagging, described through pictorial representation in Figure 15.

Overtone: A chemical molecule is propelled from the ground state to the 2nd, 3rd, or even 4th vibrational excited state when it absorbs electromagnetic radiation in the infrared region. These bands have relatively low intensity. It aids in the identification of aromatic chemicals.

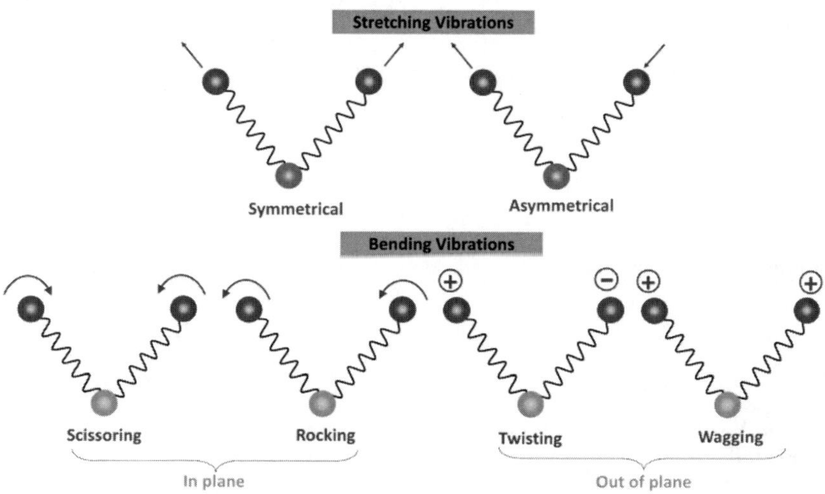

Figure 15. Major vibrational forms of stretching and bending.

Combination bands: Two fundamental vibrational frequencies (v1 + v2) pair to create a new vibrational frequency within the molecule.

Coupled Vibrations

In groups like CH_2, NH_2, and others, the linked vibrations are seen. Identical atoms join the core atom in these groupings. Because of internal disturbance, Energy from one C-H bond is moved to a neighbouring C-H bond, increasing the vibrational Frequency of $-CH_2$, which is caused by the absorption of IR radiation. Two stretching vibrational frequencies, at 2950 cm^{-1} (asymmetric stretching) and 2860 cm^{-1} (symmetric stretching), are consequently displayed by the $-CH_2$ group.

Fermi Resonance

Fermi resonance is a phenomenon in molecular spectroscopy where two or more vibrational modes in a molecule interact with each other due to their similar frequencies. This interaction leads to a splitting or shifting of the vibrational energy levels, resulting in complex spectral features. Fermi resonance often occurs in molecules with double bonds or functional groups, such as carbonyl groups.

Ex: -C=O stretching vibration was detected in benzoyl chloride at 1790 cm^{-1} & 1745 cm^{-1}. The basic vibration of C=O stretching & its fundamental overtone of C-H vibration (bending) at 875 cm^{-1} combine to produce the lower frequency band at 1745 cm^{-1} (Table 5, Figure 16).

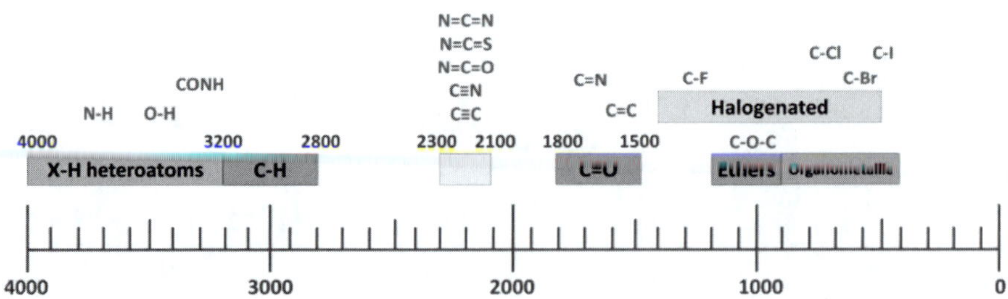

Figure 16. Pictorial representation of IR frequency correlations with functional groups.

Three sample presentation methods are frequently used in IR (Figure 17). (Duarte, 2021) (Duarte, 2021):

1. Transmission
2. Attenuated total reflection (ATR)
3. Diffuse reflectance (DRIFT)

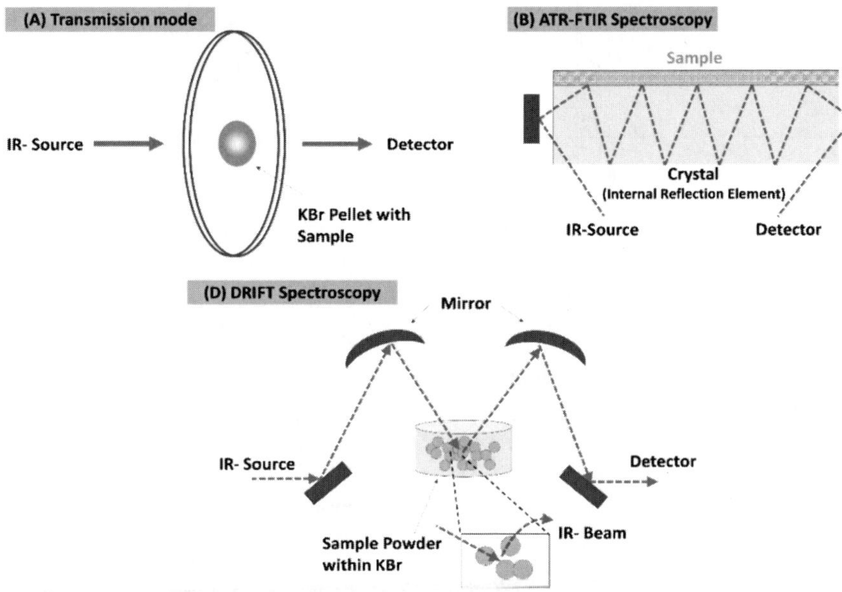

Figure 17. Different sample presentation modes in IR.

Reflection techniques, such as attenuated total reflectance (ATR), also known as multiple internal reflectance (MIR), and diffuse reflectance by Fourier transform infrared spectroscopy (DRIFT), are based on the reflection of the IR beam after it contacts the sample. As a benefit, these reflectance techniques do not require the standard sample preparation typically done for transmission FTIR spectroscopy, as they work with different optical properties (Blum and John, 2012).

Advantages and applications of IR spectroscopy:

- Organic compound identification: Due to its distinctive characteristic, IR spectroscopy facilitates the identification of many organic molecules. IR spectroscopy is used to identify compounds such as hydrocarbons, including alkenes, cycloalkanes, and aromatic hydrocarbons.

Table 5. IR frequency in relation to the functional group and type of vibrations

Functional groups	Type of vibrations	Frequency (cm^{-1})
Alkane	C-C stretching	1200
	C-H stretching	3000-2840
	-CH2 bending	1465
	-CH3 bending	1375
	CH2 rocking	720
Alkenes	C=C stretching	1650
	=C-H	3095-3010
	stretching =C-H bending	1000-650
Alkynes	C≡C stretching	2100
	≡C-H	3300
	stretching ≡C-H bending	700-600
Aromatic	C=C stretching	1600, 1500, 1450
	=C-H stretching	3040-3010
	=C-H bending	Below 900
C=O stretching	Amide	1680
	Carboxylic acid	1710
	Ketone	1715
	Aldehyde	1725
	Ester	1735
	Acid chloride	1800
	Anhydride	1760 (I), 1810 (II)
C-O	Stretching	1300-1000
O-H	Alcohol, Phenol	
	Free	3600
	H-bonded	3400-3200
	Carboxylic acid	3400-2400
Amine	N-H stretching	3440 (as), 3350 (s)
	-N-H bending	1650-1580
	C-N stretching	1350-1000
Amide	-N-H stretching	3370 (as), 3150 (s)
	-N-H bending	1650-1560
Nitriles	C≡N	2250
Imines	C=N	1690-1640
Nitro group (NO2)	N=O	1550 (as), 1350 (s)
Alkyl halide (C-X)	C-F	1350
	C-Cl	750
	C-Br, C-I	<660

Source: https://www.lkouniv.ac.in/site/writereaddata/siteContent/202004241216240526ranvijay_engg_Infrared_Spectroscopy.pdf.

- Molecular structure determination: IR spectroscopy can be used to predict the chemical structure of unknown substances. It is carried out by evaluating the locations of the absorption bands in the electromagnetic spectrum.

- Tracking the development of reactions: Examining the spectra of small amounts of the reaction mixture periodically taken from the mixture allows for easy monitoring of the progress of a chemical reaction.
- Detection of contaminants in a compound: If the reference spectrum of the pure chemical is available, it should be possible to establish whether a given sample of a compound is pure or not.
- Organic chemistry isomeric behaviour: Identification of isomers can be done using IR spectra. It is not possible for a chemical approach to distinguish between these isomers.

Limitations of IR spectroscopy:

- It is impossible to determine the molecular weight of any substance.
- The measurement of absorbance depends on the slit width and wavelength choice due to the narrowness of the spectrum.
- Does not reveal details regarding the relative positions of a molecule's functional groups.
- Contaminants, especially in water, can introduce unwanted peaks to the data, potentially hiding crucial information. This is another disadvantage.

Immunoassays for Pesticide Detection

Chromatographic techniques used in classical tests for pesticide detection often require extensive sample preparation and concentration steps to achieve detectable levels of pesticides. Consequently, these methods may fail to detect low levels of pesticide residues, potentially leading to false-negative results (Goel, 2013). Therefore, highly sensitive instrumentation is required, such as LC-MS, GC-MS, LC-MS/MS, GC-MS/MS, and LC-HRMS, among others. These instruments are costly and require sophisticated working conditions, fine sample preparation, and highly skilled personnel. In contrast to conventional and costly methods, immunoassays have demonstrated their strength in situations involving a large number of samples and the need for onsite screening. According to Beyene et al., (2019), immunoassays are recognized as one of the most potent techniques in such cases. Therefore, Immunoassays, particularly enzyme-linked immunosorbent assay (ELISA), have emerged as powerful tools for pesticide analysis.

Enzyme-Linked Immunosorbent Assay (ELISA)

ELISA is a specific technique that leverages the interaction between antigens and antibodies, forming the basis of specific immune responses that protect the body against pathogens and toxins. This principle enables ELISA to facilitate highly sensitive and selective assessment of a wide range of molecules, including nucleic acids, proteins, peptides, hormones, secondary metabolites of plants, and pesticides. Enzyme immunoassays are commonly employed, where enzymes such as alkaline phosphatase (ALP) (Zhang et al., 2015), horseradish peroxidase (HRP) (Sun et al., 2023), and β-galactosidase (Weesner et al., 2022) are utilized as labels for antigens or antibodies. The procedure involves immobilizing the antigen in the liquid phase onto a solid surface, typically a microtiter plate made of materials like polymers of styrene and vinyl chloride or propylene. Subsequently, the antigen interacts with a specific antibody, which is identified through the use of an enzyme-labelled secondary antibody. A chromogenic substrate is used to provide colour, which indicates the presence of the antigen. ALP produces a yellow colour upon hydrolysis of p-nitrophenyl phosphate, and HRP catalyzes the conversion of chromogenic substrates into coloured products. Alternatively, chemiluminescent or fluorogenic substrates can be employed for even greater sensitivity. The enzyme-substrate reactions usually reach completion within 30-60 minutes, and the addition of an appropriate stopping solution halts the reaction. Finally, the coloured or fluorescent products are quantified using a microtiter plate reader, enabling accurate detection of the analyte.

Types of ELISA Techniques

Direct ELISA: This method involves directly immobilizing the target antigen on a solid surface, typically a microplate. Subsequently, a labelled primary antibody, specifically designed to bind to the antigen, is incubated with the immobilized antigen. The labelled antibody can be directly detected, and the strength of the resulting signal corresponds to the amount of antigen available in the sample. Although direct ELISA is a straightforward and uncomplicated method, its sensitivity may be compromised by non-specific binding occurrences. There are four major types of ELISA techniques (Figure 18).

Indirect ELISA: Indirect ELISA involves a two-step process. The antigen is immobilized in the solid phase in an indirect ELISA. Then, a primary antibody specific to the antigen is added and allowed to bind to the antigen. Subsequently, a secondary antibody labeled with an enzyme is introduced. The secondary antibody recognizes and binds to the primary antibody, amplifying the signal. Indirect ELISA offers higher sensitivity and enables signal amplification, making it a preferred choice for various applications.

Sandwich ELISA: It is employed for the detection of antigens that possess multiple epitopes. Initially, a capture antibody is immobilized on a solid surface, which then binds to the target antigen. Subsequently, any unbound components are washed away, and a detection antibody labeled with an enzyme is introduced. The detection antibody specifically binds to a different epitope on the antigen, resulting in the formation of an antigen "sandwich" between the capture and detection antibodies. This technique exhibits high specificity and sensitivity as it employs two distinct antibodies that target different regions of the antigen.

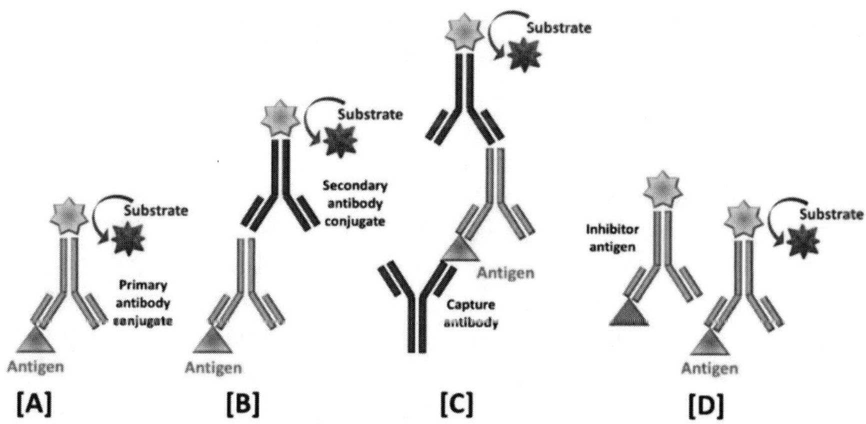

Figure 18. Types of ELISA; [A]. Direct ELISA; [B]. Indirect ELISA; [C]. Sandwich ELISA; [D]. Competitive ELISA.

Competitive ELISA: The competitive ELISA is utilized to determine the concentration of an antigen within a given sample. In this method, the sample antigen and a labelled antigen compete with each other for binding to a restricted quantity of specific antibodies that are immobilized on a solid surface. As the antigen concentration in the sample increases, the binding of the labelled antigen to the antibodies decreases proportionally. Therefore, the detected signal is inversely proportional to the concentration of the antigen within the sample. Competitive ELISA proves to be particularly valuable when measuring small molecules or accurately quantifying high concentrations of antigens.

Application of ELISA in Pesticide Detection

Over the years, researchers and scientists have explored the potential of ELISA techniques in detecting pesticide residues in various matrices, including agricultural products, soil, water, and biological samples. ELISA offers numerous advantages, such as simplicity, rapidity, cost-effectiveness, and the ability to detect multiple pesticides simultaneously. Its application in pesticide detection has played a crucial role in ensuring food safety, environmental monitoring, and regulatory compliance. With advancements in ELISA technology and the development of specific antibodies, ELISA remains a valuable tool for screening and quantifying pesticide residues, thereby contributing to the overall assessment and management of pesticide contamination. In this chapter, we discuss several recent applications of ELISA in pesticide detection.

In 2017, a new dipstick immunoassay (DIA) method using gold nanoparticles and Immunoglobulin G (IgG) bioconjugates was developed for the sensitive detection of kitazine, an organophosphorus pesticide. The DIA showed good reproducibility and accuracy in detecting kitazine residues in grape juice, highlighting its potential as a rapid qualitative screening tool for food safety (Malarkodi et al., 2017). Furthermore, commercial kit-based ELISAs detected neonicotinoid pesticides in Chinese chives using water extraction, providing accurate results (85-113% recovery). The ELISAs correlated well with high-performance liquid chromatography (HPLC), making them a practical and effective method for pesticide residue analysis in challenging matrices, such as Chinese chives (Watanabe and Miyake, 2018).

Furthermore, researchers developed an innovative technique called aptamer-based enzyme-linked immunosorbent assay (apt-ELISA), which utilizes enzyme signal amplification, to identify isocarbophos residues. This method exhibited remarkable sensitivity, surpassing the traditional ELISA, with a limit of detection as low as 0.05 ng/mL. The average recovery rate in water samples ranged from 84% to 97%, accompanied by a coefficient of variation of 4.6% to 12.1%. These findings demonstrate the suitability of the apt-ELISA approach for routine screening of isocarbophos residues (Xiang et al., 2019). In a separate study, scientists developed an in-house indirect competitive enzyme-linked immunosorbent assay (ic-ELISA) to detect chlorpyrifos residues in vegetables. The ic-ELISA exhibited good sensitivity, with an IC50 value of 0.80 μg/kg, and demonstrated low cross-reactivity. Out of 160 samples collected from northern Thailand, 33.8% tested positive for chlorpyrifos, with the highest levels detected in cucumbers, coriander, and morning glory. Chinese cabbage, cucumber, and Chinese Kale had the highest median levels. The ic-ELISA presents a rapid and effective method for quantifying chlorpyrifos residues (Hongsibsong et al., 2020).

In another investigation, nanobody-based immunoassays were developed for the rapid detection of 3-phenoxybenzoic acid (3-PBA). These included a rapid flow-through dot enzyme-linked immunosorbent assay (dot ELISA) and a gold nanoparticle (GNP) lateral-flow immunoassay. The flow-through dot ELISA achieved a limit of detection (LOD) of 0.01 ng mL^{-1}. In comparison, the GNP lateral-flow immunoassay reached an LOD of 0.1 ng mL^{-1}. Both immunoassays exhibited strong agreement with LC-MS results, confirming their suitability for rapid 3-PBA detection (Zhang et al., 2021). Moreover, in 2022, researchers conducted a study to develop an ELISA method specifically for detecting chlordimeform in aquaculture and fish farming. They prepared a polyclonal antibody against chlordimeform using a chlordimeform-BSA immunogen. The established indirect competitive ELISA demonstrated a working range of 1-20 ng/mL, an IC_{50} of 3.126 ng/mL, and a LOD of 0.637 ng/mL for chlordimeform detection. The recovery rates from spiked water samples ranged from 81% to 107%, indicating the reliability of this method for chlordimeform detection in aquaculture settings (Hu et al., 2022).

More recently, a high-affinity monoclonal antibody (mAb) was utilized to develop an ic-ELISA and colloidal gold immunochromatographic (ICA) strip for endosulfan detection. The ic-ELISA displayed an IC50 of 5.16 ng/mL and a LOD of 1.14 ng/mL. The average recoveries in spiked pear and apple samples ranged from 91.48% to 113.45%, with a coefficient of variation (CV) below 7%. For rapid visual detection, the colloidal gold ICA strip demonstrated a visual limit of detection (vLOD) of 40 ng/mL in both pear and apple samples (Zhou et al., 2023).

Benefits and Limits of ELISA Techniques

Benefits

- Simple procedure: ELISA in pesticide detection provides a straightforward and uncomplicated process, enabling ease of use and minimizing the need for extensive technical expertise.
- Easy to perform with a simple procedure, ELISA offers a user-friendly approach to pesticide detection, making it accessible to both experienced researchers and beginners due to its straightforward protocol.
- High specificity and sensitivity: ELISA exhibits a remarkable ability to accurately detect and quantify pesticides, along with high specificity for the target analytes and its sensitivity to detect even low concentrations of pesticides.
- ELISA is based on an antigen-antibody reaction: ELISA relies on the interaction between antigens (pesticides) and specific antibodies, enabling the precise identification and quantification of pesticide residues.
- High efficiency: ELISA offers a time-efficient solution for pesticide analysis, delivering rapid results and enabling high-throughput testing without sacrificing accuracy or precision.
- There is no need for complex sample pretreatment when conducting simultaneous analysis, as ELISA enables the detection of multiple pesticides in a single analysis, thereby eliminating the requirement for intricate sample preparation techniques and reducing the overall time and resources required.
- Generally safe and eco-friendly: ELISA is a safe and environmentally friendly method for pesticide detection, as it eliminates the use of radioactive substances and reduces the reliance on large quantities of organic solvents, minimizing potential hazards and promoting sustainable practices.
- Radioactive substances and large amounts of organic solvents are not required: Unlike some other detection methods, ELISA does not require the use of radioactive materials or extensive organic solvents, making it safer, easier to handle, and more environmentally friendly.
- Cost-effective assay: ELISA offers a cost-effective solution for pesticide detection, as the reagents and equipment required for the

assay are relatively affordable compared to alternative methods, reducing overall testing expenses.

- Reagents are relatively low-cost: The reagents used in ELISA for pesticide detection are relatively inexpensive, contributing to the assay's cost-effectiveness and making it a viable option for routine pesticide analysis in various settings.

Limitations

- Antibody preparation can be labour-intensive and expensive.
- ELISA requires sophisticated techniques and costly culture media.
- There is a high likelihood of encountering false-positive or false-negative results.
- Inadequate blocking of the immobilized antigen can lead to false outcomes.
- Antibodies may be prone to instability.
- Refrigerated transport and storage are necessary for antibodies due to their protein nature.

Electrochemical sensors: Chromatographic techniques such as HPLC, GC, LC-MS/MS, and GC-MS/MS are the popular techniques that are most favoured in the detection and quantification of pesticides in biotic as well as abiotic components of the environment. These techniques are highly accurate, selective, and precise; however, they are also costly and time-consuming to implement. Water, an abiotic component of the ecosystem, is highly dynamic, and contaminants present in it can move from one place to another at a rapid rate, especially in surface waters. This makes the sampling procedures difficult and may pose a potential hazard to the water for consumers. Because of this, an automatic continuous onsite measurement of the contaminant may be the solution that can provide a more accurate and thorough picture of the actual contaminant situation in the field (Wang, 2002; Wang and Chen, 1995). In this regard, electrochemical sensing technology is an alternative technology that can solve such problems as per the definition of chemical sensors given by IUPAC: "A device that converts chemical data, ranging from the concentration of a single sample component to complete composition analysis, into an analytically usable signal" (Figure 19).

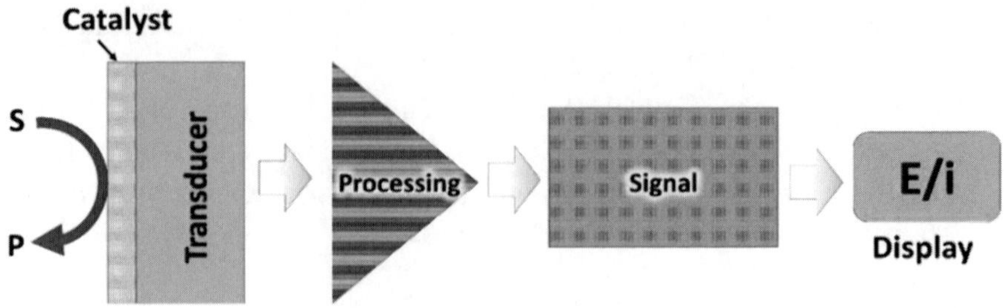

Figure 19. Sketch diagram of a chemical sensor.

Electrochemical sensing relies on measuring the electrical properties, such as charge, potential, or current, produced by the interaction of targeted analytes with sensing electrodes. A typical electrochemical cell is composed of three electrodes. They are the reference electrode, counter electrode, and working electrode. In voltammetric sensors, the electrochemical reaction primarily occurs at the working electrode, whose potential is monitored in comparison with the reference electrode. The current generated between the working electrode and the counter electrode is then recorded (Wang, 2006; Brad, 2001). Electrochemical sensors were first introduced in the 1960s by Clark and Lyons for the estimation of glucose in blood using electrodes that are responsible for sensing oxygen (Clark and Lyons, 1962).

Types of Electrochemical Sensors

Majorly, electrochemical sensors are divided into four types

a. Conductimetric electrochemical sensors
b. Potentiometric electrochemical sensors
c. Amperometric electrochemical sensors
d. Biosensors

Among all four sensors, biosensors are considered the most important tool in various scientific applications for detecting and quantifying many biochemical reactions (Anonymous, 2023).

Application of Electrochemical sensing: There are several applications of electrochemical sensing. Some of them are depicted in Figure 20.

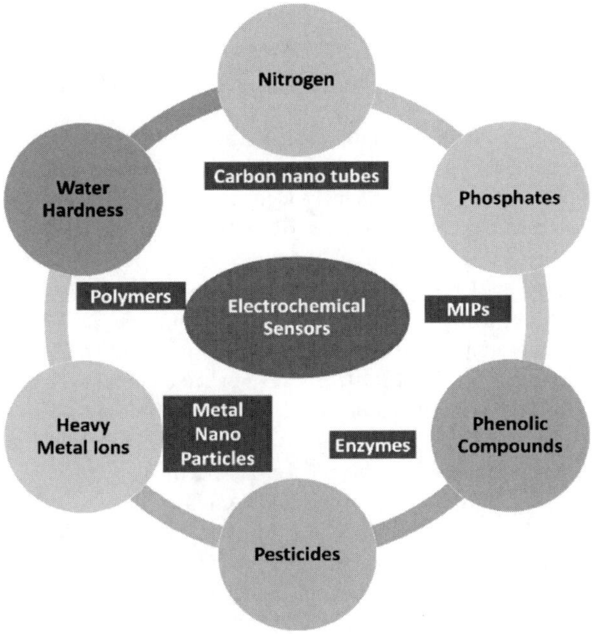

Figure 20. Applications of electrochemical sensing.

*Advantages of Electrochemical Sensors: There are several advantages to using electrochemical sensors compared to tradition*al detection and quantification techniques. Electrochemical sensors utilize electrons for signal acquisition, which is considered a clean model for analytical operations. They do not generate any waste and minimize the use of onsite portable devices. They analyze with micro volumes of the analyte. This technique has a low production cost and fast analysis, making the instrument the most favoured, leading to its popularization (Commercial glucose sensors). Furthermore, the advancement of other methods, such as chromatographic detectors, is facilitated by the development of electrochemical sensors. Electrochemical sensors, which are closely related to nanotechnology, have become more precise, selective, specific, and sensitive.

References

A Jallow, M. F., Awadh, D. G., Albaho, M. S., Devi, V. Y., Ahmad, N., 2017. Monitoring of Pesticide Residues in Commonly Used Fruits and Vegetables in Kuwait. https://doi.org/10.3390/ijerph14080833.

Ahuja, S., Scypinski, S., 2001. *Handbook of modern pharmaceutical analysis* 566.

Anonymous, 2023, *Electrochemical sensors and its applications*. Available from: https://www.researchgate.net/publication/353289661_Electrochemical_sensors_and_its_applications.

Bakirci, G. T., Hişil, Y., 2012. Fast and simple extraction of pesticide residues in selected fruits and vegetables using tetrafluoromethane and toluene followed by ultrahigh-performance liquid chromatography/tandem mass spectrometry. *Food Chem.* 135, 1901–1913. https://doi.org/10.1016/J.FOODCHEM.2012.06.051.

Bard, A. J., Faulkner, L. R. 2001. *Electrochemical Methods: Fundamentals and Applications*; Wiley: Hoboken, NJ, USA, ISBN 9780471043720.

Beyene, A. M., Du, X., Schrunk, D., Ensley, S., Rumbeiha, W. K., 2019. High-performance liquid chromatography and Enzyme-Linked Immunosorbent Assay techniques for detection and quantification of aflatoxin B1 in feed samples: A comparative study. *BMC Res. Notes* 12, 1–6. https://doi.org/10.1186/S13104-019-4538-Z/FIGURES/3.

Blum, M. M., John, H., 2012. Historical perspective and modern applications of Attenuated Total Reflectance-Fourier Transform Infrared Spectroscopy (ATR-FTIR). *Drug Test. Anal.* 4, 298–302. https://doi.org/10.1002/DTA.374.

Cavalcante, R. M., De Andrade, M. V. F., Marins, R. V, Oliveira, L. D. M., 2010. Development of a headspace-gas chromatography (HS-GC-PID-FID) method for the determination of VOCs in environmental aqueous matrices: Optimization, verification and elimination of matrix effect and VOC distribution on the Fortaleza Coast, Brazil. https://doi.org/10.1016/j.microc.2010.05.014.

Christia, C., Bizani, E., Christophoridis, C., Fytianos, K., 2015. Pesticide residues in fruit samples: comparison of different QuEChERS methods using liquid chromatography-tandem mass spectrometry. *Environ. Sci. Pollut. Res.* 22, 13167–13178. https://doi.org/10.1007/s11356-015-4456-0.

Clark, L. C., Lyons, C., 1962. Electrode systems for continuous monitoring in cardiovascular surgery. *Ann. N. Y. Acad. Sci.* 102, 29–45. https://doi.org/10.1111/J.1749-6632.1962.TB13623.X.

Cornelis, R., Caruso, J., Crews, H., Heumann, K., 2003. *Handbook of elemental speciation: techniques and methodology*.

Crnogorac, G., Schmauder, S., Schwack, W., 2008. Trace analysis of dithiocarbamate fungicide residues on fruits and vegetables by hydrophilic interaction liquid chromatography/tandem mass spectrometry. *Rapid Commun. Mass Spectrom.* 22, 2539–2546. https://doi.org/10.1002/RCM.3646.

Dąbrowski, Ł., 2018. Multidetector systems in gas chromatography. *TrAC - Trends Anal. Chem.* https://doi.org/10.1016/j.trac.2018.02.006.

Duarte, E. M. and N., 2021. Mid-Infrared Spectroscopy as a Valuable Tool to Tackle Food.

Dzuman, Z., Zachariasova, M., Veprikova, Z., Godula, M., Hajslova, J., 2015. Multi-analyte high performance liquid chromatography coupled to high resolution tandem mass spectrometry method for control of pesticide residues, mycotoxins, and pyrrolizidine alkaloids. *Anal. Chim. Acta* 863, 29–40. https://doi.org/10.1016/J.ACA.2015.01.021.

Ettre, L. S., 2003. The beginnings of gas chromatography and early symposia held in the United States. *LC-GC North Am.* 21, 900–907.

Goel, P., 2013. Immunodiagnosis of pesticides: A review. *African J. Biotechnol.* 12, 7158–7167. https://doi.org/10.4314/AJB.V12I52.

Hamilton, R. J., Sewell, P. A., 1982. *Introduction to high performance liquid chromatography. Introd. to high Perform. Liq. Chromatogr.* 1–12. https://doi.org/10.1007/978-94-009-5938-5_1.

Hongsibsong, S., Prapamontol, T., Xu, T., Hammock, B. D., Wang, H., Chen, Z. J., Xu, Z. L., 2020. Monitoring of the Organophosphate Pesticide Chlorpyrifos in Vegetable Samples from Local Markets in Northern Thailand by Developed Immunoassay. *Int. J. Environ. Res. Public Heal.* 2020, Vol. 17, Page 4723 17, 4723. https://doi.org/10.3390/IJERPH17134723.

Hu, Y. mei qing, Feng, H. wei, Liu, S., Liu, C., Zhao, P. yan, Zhang, M., Zhang, L., Zhao, J., Li, J. zeng, Yu, X. miao, Liu, H. sheng, Zhu, J. feng, 2022. The preparation of polyclonal antibody against chlordimeform and establishment of detection by indirect competitive ELISA. 57, 114–124. https://doi.org/10.1080/03601234.2022.2029274.

Jennings, W., Mehran, M. F., 1986. Sample Injection in Gas Chromatography. *J. Chromatogr. Sci.* 24, 34–40. https://doi.org/10.1093/chromsci/24.1.34.

Joshi, V. S., Kumar, V., Rathore, A. S., 2015. Role of Organic Modifier and Gradient Shape in RP-HPLC Separation: Analysis of GCSF Variants. *J. Chromatogr. Sci.* 53, 417–423. https://doi.org/10.1093/chromsci/bmu222.

Jumhawan, U., Bamba, T., 2010. Supercritical Fluid Chromatography. *Appl. Green Solvents Sep. Process.* 82, 483–516. https://doi.org/10.1016/B978-0-12-805297-6.00016-4.

Łozowicka, B., Rutkowska, E., Jankowska, M., n.d. Influence of QuEChERS modifications on recovery and matrix effect during the multi-residue pesticide analysis in soil by GC/MS/MS and GC/ECD/NPD. https://doi.org/10.1007/s11356-016-8334-1.

Malarkodi, C., Rajeshkumar, S., Annadurai, G., 2017. Detection of environmentally hazardous pesticide in fruit and vegetable samples using gold nanoparticles. *Food Control* 80, 11–18. https://doi.org/10.1016/J.FOODCONT.2017.04.023.

Miao, S., Wei, Y., Pan, Y., Wang, Y., Wei, X., 2023. Detection methods, migration patterns, and health effects of pesticide residues in tea. *Compr. Rev. Food Sci. Food Saf.* 1–32. https://doi.org/10.1111/1541-4337.13167.

Michalke, B., 2002. The coupling of LC to ICP-MS in element speciation: I. General aspects. *TrAC Trends Anal. Chem.* 21, 142–153. https://doi.org/10.1016/S0165-9936(01)00146-7.

Passos, L. C., M., M. F.S., Saraiva, M. L., 2019. Detection in UV-visible spectrophotometry: Detectors, detection systems, and detection strategies. *Meas. J. Int. Meas. Confed.* 135, 896–904. https://doi.org/10.1016/j.measurement.2018.12.045.

Rahman, M. M., Abd El-Aty, A. M., Choi, J.-H., Shin, H.-C., Shin, S. C., Shim, J.-H., 2015. Basic Overview on Gas Chromatography Columns. *Anal. Sep. Sci.* 823–834. https://doi.org/10.1002/9783527678129.assep024.

Reichenbach, S. E., Rempe, D. W., Tao, Q., Bressanello, D., Liberto, E., Bicchi, C., Balducci, S., Cordero, C., 2015. Alignment for Comprehensive Two-Dimensional Gas Chromatography with Dual Secondary Columns and Detectors. *Anal. Chem.* 87,

10056–10063. https://doi.org/10.1021/ACS.ANALCHEM.5B02718/SUPPL_FILE/AC5B02718_SI_001.PDF.

Reinecke, M. G., 2004. Organic Spectroscopic Analysis By R. J. Anderson, D. J. Bendell, and P. W. Groundwater (University of Sunderland). Royal Society of Chemistry, Cambridge. 2004. vi + 176 pp. 19 × 25 cm. $27.25. ISBN 0-85404-476-0. *J. Nat. Prod.* 67, 2158–2158. https://doi.org/10.1021/NP030773B.

Rong, L., Wu, X., Xu, J., Dong, F., Liu, X., Pan, X., Du, P., Wei, D., Zheng, Y., 2018. Simultaneous determination of three pesticides and their metabolites in unprocessed foods using ultraperformance liquid chromatography-tandem mass spectrometry. *Food Addit. Contam. - Part A Chem. Anal. Control. Expo. Risk Assess.* 35, 273–281. https://doi.org/10.1080/19440049.2017.1398419.

Saha, C., Bhushan, M., Singh, L. R., 2023. Pesticide sensing using electrochemical techniques: a comprehensive review. *J. Iran. Chem. Soc.* 20, 243–256. https://doi.org/10.1007/S13738-022-02666-W/METRICS.

Sindhu, S., Manickavasagan, A., 2023. Nondestructive testing methods for pesticide residue in food commodities: A review. *Compr. Rev. Food Sci. Food Saf.* 22, 1226–1256. https://doi.org/10.1111/1541-4337.13109.

Sun, Y., Wen, L., Ma, H., Ma, W., Fu, Z., Li, Y., Zhang, C., Li, L., Liu, J., 2023. Engineering trienzyme cascade-triggered fluorescent immunosensor platform by sequentially integrating alkaline phosphatase, tyrosinase and horseradish peroxidase. *Chinese Chem. Lett.* 34, 107654. https://doi.org/10.1016/J.CCLET.2022.06.077 uv vis iupac 2002, 2002. in: *Methods in Geochemistry and Geophysics*. pp. 48–88. https://doi.org/10.1016/B978-0-444-40826-6.50008-X.

Valkó, K., 2000. *Separation methods in drug synthesis and purification* 589. Wang, J., 2002. Real-Time Electrochemical Monitoring: Toward Green Analytical Chemistry. *Acc. Chem. Res.* 35, 811–816. https://doi.org/10.1021/AR010066E.

Wang, J. 2006. *Analytical Electrochemistry*, 3rd ed.; John Wiley & Sons, Inc.: Hoboken, NJ, USA, ISBN 9780471678793.

Wang, J., Chen, Q., 1995. Remote electrochemical biosensor for field monitoring of phenolic compounds. *Anal. Chim. Acta* 312, 39–44. https://doi.org/10.1016/0003-2670(95)00207-G.

Wang, T., 2007. Liquid Chromatography–Inductively Coupled Plasma Mass Spectrometry (LC–ICP–MS). 30, 807–831. https://doi.org/10.1080/10826070701191094.

Watanabe, E., Miyake, S., 2018. Direct determination of neonicotinoid insecticides in an analytically challenging crop such as Chinese chives using selective ELISAs. 53, 707–712. https://doi.org/10.1080/03601234.2018.1480154.

Weesner, J. A., Annunziata, I., Yang, T., Acosta, W., Gomero, E., Hu, H., van de Vlekkert, D., Ayala, J., Qiu, X., Fremuth, L. E., Radin, D. N., Cramer, C. L., d'Azzo, A., 2022. Preclinical Enzyme Replacement Therapy with a Recombinant β-Galactosidase-Lectin Fusion for CNS Delivery and Treatment of GM1-Gangliosidosis. *Cells* 11, 2579. https://doi.org/10.3390/CELLS11162579/S1

Williams, T. I., 1979. 75 Years of chromatography — A historical dialogue. *Endeavour* 3, 135–136. https://doi.org/10.1016/0160-9327(79)90124-8.

Xiang, L., Wu, H., Cui, Z., Tang, J., 2019. Indirect Competitive Aptamer-Based Enzyme-Linked Immunosorbent Assay (apt-ELISA) for the Specific and Sensitive Detection of Isocarbophos Residues. 52, 1966–1975. https://doi.org/10.1080/00032719.2019.1587446.

Zhang, C., Wu, X., Li, D., Hu, J., Wan, D., Zhang, Z., Hammock, B. D., 2021. Development of nanobody-based flow-through dot ELISA and lateral-flow immunoassay for rapid detection of 3-phenoxybenzoic acid. *Anal. Methods* 13, 1757–1765. https://doi.org/10.1039/D1AY00129A.

Zhang, Z., Chen, Z., Wang, S., Cheng, F., Chen, L., 2015. Iodine-Mediated Etching of Gold Nanorods for Plasmonic ELISA Based on Colorimetric Detection of Alkaline Phosphatase. *ACS Appl. Mater. Interfaces* 7, 27639–27645. https://doi.org/10.1021/ACSAMI.5B07344/SUPPL_FILE/AM5B07344_SI_001.PDF.

Zhou, X., Guan, S., Li, N., Zhou, J., Pan, W., Wang, Y., 2023. Development of Indirect Competitive ELISA and Colloidal Gold Immunochromatographic Strip for Endosulfan Detection Based on a Monoclonal Antibody. *Foods* 2023, Vol. 12, Page 736 12, 736. https://doi.org/10.3390/FOODS12040736.

Zwinkels, J., 2020. *Electromagnetic Spectrum*. Springer Berlin Heidelberg, Berlin, Heidelberg. https://doi.org/10.1007/978-3-642-27851-8.

Chapter 3

Insecticide Detection in the Environment: Recent Advances and Prospects

Indu Chopra[1,*]
Ankita Trivedi[1]
Reena Chauhan[2]
and Sharat Kothari[3]
[1]Department of Soil Science and Agricultural Chemistry, ICAR-IARI, New Delhi, India
[2]Department of Entomology, CCS HAU, Hisar, India
[3]Forest Ecology and Climate Change Division, ICFRE-Arid Forest Research Institute, Jodhpur, Rajasthan, India

Abstract

As there is a continuous demand for food, to reduce crop losses and sustain agricultural production, the application of insecticides and pesticides in agricultural fields is inevitable. However, the continued and non-thoughtful use of insecticides has resulted in the accumulation of harmful residues, including metabolites and the degraded products of the parent molecule in water, soil, plants, and animal products. In order to monitor and assess the risks associated with environmental exposure and generate information on the concentration of insecticides and their residues at the target site, detection and quantification of these insecticides needs to be done at regular intervals that can help not only to know the level of contamination at a particular site but also to increase the sensitivity of the already developed methods by necessary modifications and develop economically viable techniques by

* Corresponding Author's Email: tinaindu@gmail.com

In: Insecticides
Editors: Archana Anokhe and Govindaraj Kamalam Dinesh
ISBN: 979-8-90134-022-6

developing or identifying/selecting cost-effective materials for the on-site detection and quantification of insecticides with enhanced accuracy and sensitivity. The objective of this chapter is to discuss and describe the analytical methods being used to detect insecticides in different environmental components and food commodities. Also, a brief overview of insecticide classification, sampling, available extraction methods, and recent advancements in their detection has been provided to have an overall understanding of insecticide detection.

Keywords: insecticides, detections, environments, methods

Introduction

As the lifestyle and purchasing power of the common person have changed, consumer awareness and inclination towards selecting quality products have also increased. Earlier, fruits and vegetables were only available during their season, but due to advancements in varieties, technologies, farming practices, and storage facilities, all fruits and vegetables are now available year-round, regardless of the season. To ensure these fruits and vegetables are available throughout the year and remain safe from insect and pest infestations, the reliance on crop protection chemicals, such as insecticides and fungicides, has also increased. Besides the quality and quantity of the crop produced in terms of crop production and crop productivity, these aspects also need to be maintained or enhanced to meet the food requirements of the burgeoning population. According to FAO, there is a need to increase food production by 80% in developing countries to keep pace with the swelling population. To meet the continuously rising food demand, reduce crop losses, and sustain agricultural production, the application of insecticides and pesticides in agricultural fields is inevitable (Petersen et al., 2019). Insecticides are the substances/ mixtures that are used for the prevention of the destruction of insects (human and animal disease vectors included) that can hinder food production, processing, or storage. Due to these properties, insecticides and pesticides are widely used to control pests and thereby enhance crop production. Globally, from 1990 to 2010, the average usage of insecticides, herbicides, and fungicides was 342,000, 566,000, and 353,000 tons, respectively (Y. Liu et al., 2015).

Additionally, Bernhardt et al., (2017) reported that global pesticide usage is estimated to be approximately 6 million metric tons. In developing countries, although pesticide application is 25%, it is exceptionally higher in vegetables (de Bon et al., 2014). However, the continued and non-thoughtful use of insecticides has resulted in the accumulation of harmful residues, including metabolites and the degraded products of the parent molecule in water, soil, plant, and animal products (Topuz et al., 2005). It has also led to resistance to these crop protection chemicals and a resurgence of the next-gen insect population. Prolonged use of these chemicals has also led to reduced effectiveness in controlling insect-pest attacks, as well as the need for higher doses of these insecticides (Dusfour et al., 2019). The persistence of insecticides and their metabolites has resulted in residues of these chemicals in various matrices, posing health hazards to humans and other living organisms. To monitor and assess the risks associated with environmental exposure and generate information on the concentration of insecticides and their residues at the target site, detection and quantification of these insecticides need to be done at regular intervals (N. Kumar et al., 2019).

Monitoring of the insecticides in consumables can help not only to know the level of contamination at a particular site but also to increase the sensitivity of the already developed methods by necessary modifications and develop economically viable techniques by developing or identifying/selecting cost-effective materials for the on-site detection and quantification of insecticides with enhanced accuracy and sensitivity (Sulaiman et al., 2019). Constant monitoring of insecticide residues in food and environmental samples is also essential to ensure that their levels comply with legal restrictions, allowing for the uninterrupted export of commodities without compromising their quality. In this context, several analytical methods have been developed and used to detect the residues of parent compounds and their metabolites, such as Mass spectroscopy (MS), Gas chromatography (GC), High-Performance Liquid Chromatography (HPLC) (Samal et al., 2023). Recently, several advances have also been made in existing techniques, in addition to the development of new methods and methodologies to identify and quantify insecticides and their residues at ppb levels in various matrices. Nanotechnology-based technologies have also been explored and employed for the detection/ analysis of various insecticides. Sensor-based techniques have been developed and utilized for on-site monitoring and real-time analysis of various compounds and their residues, offering simplicity of operation while maintaining high sensitivity and selectivity.

Additionally, efforts are ongoing to develop economical, sustainable, sensitive, reliable, accurate, and user-friendly analytical methods that can help identify residues of insecticides in food commodities and agricultural produce. Therefore, this chapter is intended to describe both traditional techniques and recent developments in the field of insecticide detection and quantification. The objective of this chapter is to discuss and describe the analytical methods being used to detect insecticides in different environmental components and food commodities. Also, a brief overview of insecticide classification, sampling, available extraction methods and recent advancements in their detection has been provided to have an overall understanding of insecticide detection.

Insecticides

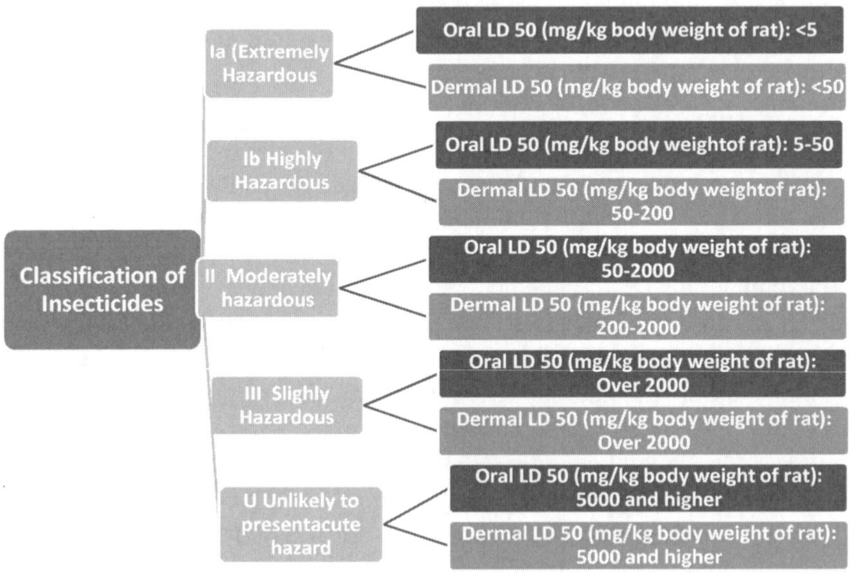

Figure 1. WHO Classification of Insecticides Based on Acute Toxicity Hazard.

Insecticides are chemical agents intended to kill or prevent insects and their destructive behavior. These compounds can be used both on/ off the field (in storage) to prevent damage caused by insects. Based on their origin, insecticides can be classified into two categories: natural and synthetic (R. Kaur et al., 2019). The insecticides of synthetic origin are further classified into organic and inorganic classes. The inorganic insecticides primarily comprise metal-based, non-carbonaceous compounds that are highly effective but non-biodegradable, posing environmental problems (Tudi et al., 2021). The synthetic organics are further sub-classified into organophosphates, carbamates, organochlorines, pyrethroids, and sulfonylureas. These compounds are widely used in agriculture globally to enhance crop production and productivity by controlling insect attacks on various crops, as recommended. Insecticides are also classified based on their chemistry, toxicity, mode of action, and absorption or penetration (Tudi et al., 2021). The insecticides have also been classified by WHO using Acute Toxicity Hazard categories based on the oral and dermal routes that have been defined through lethal dose (LD_{50}, insecticide dose required to kill half the insect population under study applied through dermal or oral digestion) under Globally Harmonized System (GHS) (World Health Organization, 2020) in consultation with different countries, various international agencies and regional bodies. The classification of insecticides according to the WHO, aligned with the GHS categories, is given in Figure 1.

Detection of Insecticides and Their Residues in Environmental Matrices

The analysis of trace insecticides in food commodities and various environmental matrices has become essential, as insecticides are commonly applied to enhance crop yield and quality. However, due to the diverse nature and structure of insecticides applied and the complexity of the matrix, residue detection can be challenging and thus requires different pre-treatment methodologies to enhance method accuracy and sensitivity in determining residues of various commodities. The analytical determination of insecticides and their residues in various commodities involves four steps, primarily as outlined in Figure 2.

Preparation of Samples

Sample preparation is the most crucial and arduous step on which the whole analytical task depends. It is a crucial part of any analysis, as even the slightest alteration in the form of the analyte or loss in quantity can compromise the validity of analytical results due to changes in accuracy, precision, and recovery. For adequate sampling, the selection of the number of sites, the number of samples, the sampling technique (random or systematic), the number of replications, the time of sampling, storage, and record maintenance needs to be considered for an accurate estimation of the analytes under study. To prepare the samples effectively, the physicochemical properties of the analyte to be determined, as well as the sample matrix, must be part of the experimental planning.

Extraction/Clean-Up of the Samples

To eliminate interferences from different matrices and determine the accurate amount of insecticide in a particular commodity, it is crucial to clean the samples carefully. For isolating toxicologically significant analytes and their metabolites after removal of unwanted impurities from a particular matrix, extraction/clean-up is performed using physical or chemical methods.

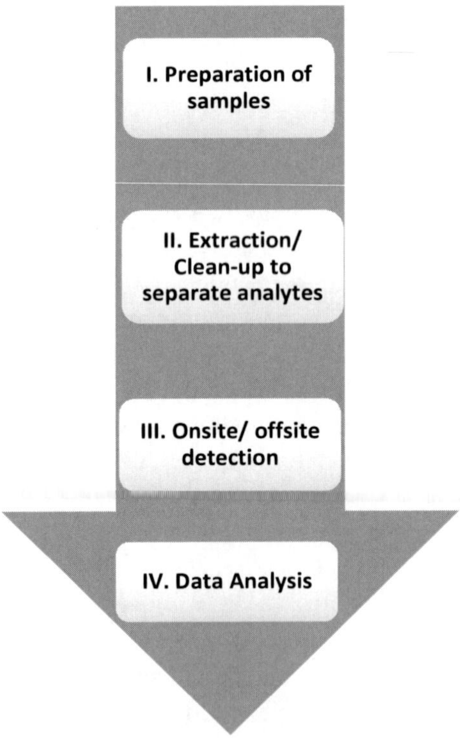

Figure 2. Steps for analytical determination of insecticides and their residues.

Table 1. Summary of different extraction techniques for insecticides

S No.	Name of the Technique	Application	Benefits	Drawbacks
1	Liquid-Liquid Extraction (LLE)/ Solvent Extraction	The analyte is dissolved in two immiscible liquids/ solvents	Consistent and easy to handle, effective for different analytes.	Large volume of toxic solvents needed, Arduous technique
2	Solid phase extraction (SPE)	Involves analyte adsorption on solid sorbent materials	Reduced time-consumption in comparison to LLE, powerful technology to extract and pre-concentrate	Pre-treatment is required, hazardous organic Solvents are used
3	Quick, Easy, Cheap, Rugged, Effective and Safe (QuEChERS)	Used for multi-residue analysis. Polar and non-polar analytes can be extracted simultaneously.	A wide range of analytes can be extracted using an efficient and straightforward technique. A low volume of hazardous solvents is required	Enrichment factors are low, Modification in the procedure is needed
4	Solid phase micro-extraction (SPME)	The target analyte is distributed between the matrix and the solid stationary phase through micro-extraction	Simple, fast, and easy to handle, Low solvent volumes needed	Sample carry-over is difficult, Limited lifetime because of the fibre coating
5	Dispersive liquid-liquid micro-extraction (DLLME)	Dispersive solvents miscible with both water and extraction solvent are used for organic/ inorganic analytes from aqueous samples	Easy to handle, Nominal use of hazardous solvents, Extraction speed is good, and low cost	Extraction efficiency low
6	Single drop micro-extraction (SDME)	Samples can be extracted by consuming a small volume of extraction solvent in the form of a drop using a microsyringe.	Rapid and economical, Easy to handle, Eco-friendly and renewable, Lower requirement for toxic solvents	The stability of the suspended drop is low, Maintenance of low volume is difficult Extraction time is long
7	Hollow fibre-liquid phase micro-extraction (HF-LPME)	Analytes are extracted from the aqueous phase to the porous fibre by diffusion	Economical, fertilization factors better, High reproducibility, and sample cleanliness ability	Full automation is complex. Sample contamination chances are high.

Table 1. (Continued)

S No.	Name of the Technique	Application	Benefits	Drawbacks
8	Continuous Flow Micro-extraction (CFME)	The extraction droplet is held continuously with the tip of the fibre tubing contained in a glass chamber, resulting in Higher analyte pre-concentration	Consumption of hazardous solvents is low. Inexpensive tools are used. Extraction from complex matrices is simple, fast, and effective.	Volume of Micro-drop is limited Insertion of the drop in the glass extraction chamber is difficult
9	Solid Liquid Extraction with low-temperature partitioning (SLE/LTP)	The analyte is partitioned between the aqueous phase and organic solvent at a lower temperature (-20°C), and the analyte in the organic phase is isolated	One-step extraction: No need for clean-up as co-extractives freeze in the aqueous phase. Low solvent consumption,	Time-consuming,
10	Microwave Assisted Extraction (MAE)	Microwave energy is used to heat the analyte, extracting it into the solvent rapidly.	Less extraction time Low solvent requirement Higher solvent throughput	Additional extraction to be carried out Method development is complex

Table 2. Brief description of extraction methods and detection techniques of insecticides/pesticides in environmental matrices

Extraction method	Detection technique	Matrix	Insecticide/pesticide	LOD	Reference
Food					
Molecularly imprinted solid phase extraction	HPLC	Blank rape and cauliflower	Trichlorfon	4.2 μg g^{-1}	(X. Wang et al., 2014)
			Monocrotophos	1.2 ng g^{-1}	
Polystyrene-coated magnetic nanoparticles-based extraction	HPLC	Cabbage, pakchoi, Chinese kale, rape, Chinese chive, lettuce, amaranth, broccoli, cauliflower, Chinese cabbage	Betacyfluthrin	0.0324 ng g^{-1}	(Yu & Yang, 2017)
			Decamethrin	0.0200 ng g^{-1}	
			Fenvalerate	0.0392 ng g^{-1}	
			Permethrin	0.0384 ng g^{-1}	
			Bifenthrin	0.0288 ng g^{-1}	
Single-phase extraction	GC–QqQ-MS/MS	Orange and cucumber	Multiclass pesticide	6-8 μg kg^{-1}	(Moreno et al., 2008)
QuEChERS	GC-ECD/NPD	Tomato, apple, green bean	Multiclass pesticide	0.005-0.01 mg kg^{-1}	(Aysal et al., 2007)
-	DC-ELISA	Water, soil, lettuce, cabbage	Carbofuran	0.11 ng ml^{-1}	(Yang et al., 2008)
-	ELISA	Tangerine juice	Chlorpyrifos	0.20 ± 0.04 μg L^{-1}	(Navarro et al., 2013)
			Fenthione	0.50 ± 0.06 μg L^{-1}	
-	ELISA	Grape, Chinese cabbage, soil, water	Chlorpyrifos methyl	0.32 ng ml^{-1}	(Qian et al., 2009)
Molecularly imprinted polymers	CE-BI	Kidney bean and cucumber	Trichlorfon	0.16 mg L^{-1}	(J. Li et al., 2017a)
Liquid-liquid extraction	HPLC-MS	Fruit juices, Berry juices, Olives	Malathion Triadimefon Diazinon	3×10^{-2} to 3×10^{-4} mg L^{-1}	(Timofeeva et al., 2017)
Solid phase extraction	HPLC	Fruit juice	Flusilazole, Fenbuconazole Myclobutanil Penconazole Epoxiconazole	0.5-1 ng kg^{-1}	(G. Liu et al., 2018)

Table 2. (Continued)

Extraction method	Detection technique	Matrix	Insecticide/pesticide	LOD	Reference
QuEChERS extraction	LC-MS	Sunflower oil, tomato, spinach, lettuce	Chlordane Dicofol Dieldrin	0.001-0.011 mg kg^{-1}	(C & G V, 2017; Neufeld et al., 2015)
QuEChERS extraction with magnetic solid phase extraction and dispersive liquid-liquid microextraction	GC flame ionization detection	Vegetables, fruit, and nectar	Iazinon, Penconazole, Oxadiazon, Diniconazole, and Fenazaquin	5-200 μg L^{-1}	(Farajzadh et al., 2019)
Matrix solid-phase dispersion	GC–MS	Apples	Multiclass pesticide	-----	(Purdešová & Dömötorová, 2017)
Matrix solid-phase dispersion	LC-MS/MS	Rice	Chlorfenapyr and abamectin	0.8 ng g^{-1}	(P. Zhang et al., 2017)
Solid phase microextraction	GC-MS	Fruits, vegetables, water	Parathion, Diazinon, Chlorpyrifos, Fenthion	0.005-0.020 μg L^{-1}	(Saraji et al., 2016)
Single drop microextraction	GC flame photometric detector	Milk, honey, water	Multiclass pesticide	0.5-1 ng kg^{-1}	(Shamsipur et al., 2016)
QuEChERS extraction	GC-MS	Milk, dairy	Multi-pesticide	< 5 mg kg^{-1}	(Manav et al., 2019)
Solid phase extraction	ELISA	Chinese cabbage, apple	Phenylpyrazole, fipronil, Organophosphorus Neonicotinoid, dinotefuran	3.7-162.2 ng mL^{-1}	(Vasylieva et al., 2015)
Solid phase extraction	Surface-enhanced Raman spectroscopy	Orange, apple, tomato	Phosmet, Chlorpyrifos, Carbaryl, Phosmet, Azinphos-methyl	2.94-6.66 ppm	(B. Liu et al., 2013; Yande et al., 2016)
QuEChERS extraction	Colorimetric sensor	Vegetable	Chlorpyrifos	8.6 ppm	(Kim et al., 2018)
Solid phase extraction	Colorimetric sensor	Vegetable	Methyl paraoxon, Chlorpyrifos-oxon	5.3-18 ng mL^{-1}	(Busa et al., 2016)
Solvent extraction	Glucose meter	Apple, cucumber	Organophosphorus	5 μg L^{-1}	(Tang et al., 2019)

Extraction method	Detection technique	Matrix	Insecticide/pesticide	LOD	Reference
Direct	Colorimetric gold nanoparticles	Beer	Glyphosate	0.01 ppb	(M. J. Tan et al., 2017)
Solid phase extraction	Aptasensors	Apple	Malathion	0.5 pM	(Bala et al., 2018; Nie et al., 2018)
Direct	Aptasensors	Fruit juice	Acetamiprid	0.086 µM	(Rapini et al., 2016)
Solid phase extraction	Aptasensor	Soya milk, mango juice, tomato, plum	Carbendazim	8.2 pg mL^{-1}	(Eissa & Zourob, 2017)
Solid phase extraction	Molecular Imprinted Polymer Sensors	Water, Fruit juice, Vegetable juice	Paraoxon	$1x10^{-9}$ M	(F. R. Wang et al., 2018)
Solid phase extraction	Capillary electrophoresis	Kidney beans, Cucumber, Leek, samples	Trichlorfon, Paraquat Diqua	0.15–0.20 ng mL^{-1}	(Chui et al., 2017; J. Li et al., 2017b)
Solid phase extraction	Micellar electrokinetic capillary chromatography	Fruits, Vegetables	Azoxystrobin Kresoxim-methyl Pyraclostrobin Nicotinyl	0.001 mg kg^{-1}	(Guo et al., 2017b)

Though there are several techniques available for the extraction of the analytes and their metabolites from the samples under study before the selection of any of the techniques to extract the maximum amount of the analyte, the following points must be considered:

- Matrix from which the target analyte has to be extracted (fruit, vegetable, soil, water, *etc.*)
- Chemical class of the analyte and its thermal stability
- Precautions and safety measures
- Possible co-extractives, along with the analyte
- Selection of a suitable solvent or solvent mixture for the analyte extraction (should have high solubility for analytes and low for the co-extractives)

Extraction is a common technique used to separate target analytes and their residues or metabolites from complex sample matrices by preconcentrating the analyte(s) into a suitable solvent(s). This process enables the detection and analysis of toxic compounds, including insecticides and/or their metabolites, through a suitable detection technique. A brief overview of various extraction techniques used for the pre-concentration of target analytes (e.g., insecticides), along with their benefits and drawbacks, is presented in Table 1.

Analytical Determination and Quantification of Insecticide Residues and Data Analysis

After extracting and concentrating the target analyte in a suitable solvent, the determination of the insecticides and their residues/ metabolites is carried out. For this purpose, several techniques have been developed and applied by researchers to detect various insecticides in different environmental and food samples. Traditionally, chromatographic methods are used for the detection and quantitative analysis of insecticides. The compounds are separated between the two phases (mobile phase and stationary phase) depending on their polarity and interaction with the stationary phase. These methods include gas chromatography (GC), liquid chromatography (LC), High-performance liquid chromatography (HPLC), and Micellar electrokinetic capillary chromatography (MEKC). Additionally, the analytical determination of the insecticides has been carried out using capillary electrophoresis (CE) and enzyme-linked immunosorbent assay (ELISA).

Gas Chromatography (GC)

Gas chromatography is one of the most commonly used methods for separating analytes based on their volatilities (Bahaghighat et al., 2019). The technique applies to compounds that are non-polar, volatile, and can be vaporized efficiently. The detection of insecticides by GC is carried out by coupling it with a specific detector, depending on the properties of the insecticide being analyzed. Different detectors used in GC include the electron capture detector (ECD), flame photometric detector (FPD), nitrogen phosphorus detector (NPD), and flame ionization detector (FID). For the detection of residues of organochlorines and other halogenated compounds, ECD is used. To detect insecticides with sulfur and phosphorus groups, a flame photometric detector is used. NPD, the thermionic detector, is used for the selective detection of analytes containing nitrogen and phosphorus (Ramos et al., 2009). For the monitoring and detection of almost all groups of insecticides, FID can be used. To improve sensitivity and accuracy, a mass-selective detector has also been added to the list (Łozowicka et al., 2017). The application of tandem MS has also helped improve sensitivity and reduce interferences, enabling more effective and accurate analysis of the target analytes. The efficiency of the analysis has been increased with the use of different analyzers, including ion trap (IT) (Tao et al., 2009), quadrupole (de Oliveira et al., 2012), triple quadrupole (QqQ) (C. C. Wu, 2017), and time of flight mass analyzer (TOF) (Cervera et al., 2012).

Liquid Chromatography (LC)

The technique is used to separate and analyse the toxic compounds that are highly polar and non-volatile/ thermolabile. To detect the pesticide residues, LC is often combined with detectors such as ultraviolet (UV), fluorescence detector, photo-diode array (PDA) detector, diode-array (DAD) detectors, and mass (MS) detectors like triple quadrupole and Q-trap (Rong et al., 2017; Tadeo et al., 2010). The application of MS for residue detection is successful for both multi-residue and quantitative analysis of insecticides and pesticides in various matrices due to its high sensitivity and selectivity.

High-performance liquid chromatography (HPLC) is primarily used for the precise separation of highly polar and non-volatile compounds as the solvent travels through the column under high pressure, facilitated by a pump to prevent pressure drops, thereby reducing the separation time. It can be combined with various detectors for the efficient detection of residues in different matrices. To further reduce analysis time with enhanced chromatographic resolution and sensitivity, Ultra High-Performance Liquid Chromatography (UHPLC), an improved version of HPLC, has been reported in recent years (Bilehal et al., 2014). Sivaperumal et al., (2015) prepared the samples using solid-phase extraction cartridges. They employed ultra-high-performance liquid chromatography/time-of-flight mass spectrometry (UHPLC/TOF-MS) for the detection of 60 pesticides in vegetables.

Micellar Electrokinetic Capillary Chromatography (MEKC)

Micellar electrokinetic capillary chromatography (MEKC) is a hybrid technique that combines the principles of chromatography and electrophoresis, utilizing the separation of analytes due to differential partitioning between the two phases, namely the mobile aqueous phase and the micellar pseudo-stationary phase (Hancu et al., 2013). This technique has been able to provide a concentration factor in the range of 10–5000 without requiring any complicated clean-up or other off-line procedures to pre-concentrate the samples, thus offering a significant opportunity for residue analysis (Quirino & Terabe, 1998).

Guo et al., (2017) developed a nonaqueous micellar electrokinetic capillary chromatography method with indirect LIF for the determination of strobilurin fungicide residues in fruits and vegetables. The results showed that the detection limits of azoxystrobin, kresoxim-methyl, and pyraclostrobin were 0.001 mg/kg, their linear dynamic ranges were 0.005–2.5 mg/kg, and the recoveries of the spiked samples were 81.7–96.1%, 86.5–95.7%, and 87.3–97.4%, respectively. This method was found to be sensitive to detect these fungicides at their maximum residue limits.

Capillary Electrophoresis (CE)

This technique is one of the most striking analytical methods for the rapid analysis of toxic compounds, both qualitatively and quantitatively, with variations in polarity and molecular weight. This technique can be used for the analysis of both small molecules, such as amino acids, and large macromolecules, including proteins and nucleic acids. In this technique, a reagent is required in smaller quantities, and the analyte is separated in a short time with high separation efficiency. Li et al., (2017) developed a biomimetic immunoassay-capillary electrophoresis method with enhanced sensitivity for detecting trichlorfon in vegetables. Under optimal conditions, the limit of detection (LOD, IC_{15}) and the sensitivity (IC_{50}) of this method were 0.16 mg/L and 0.13 μg/L for trichlorfon.

Enzyme-Linked Immunosorbent Assay (ELISA)

ELISA is an analytical technique based on antigen-antibody interaction through covalent bonding, which is highly sensitive and specific in detecting insecticides and their residues. This technique is simple, inexpensive, and rapid, and can be carried out by personnel with minimal training. Navarro et al., (2013) used a rapid duplex ELISA for the simultaneous determination of the two most widely used organophosphorus insecticides in tangerine juices. The optimized duplex ELISA was accomplished within 40 minutes with detection limits of 0.20 ± 0.04 μg/L and 0.50 ± 0.06 μg/L for chlorpyrifos and fenthion, respectively. However, the stability of the isolated antibodies and insufficient blocking of immobilized antigens are the major drawbacks of this technique that can lead to false results. Table 2 provides a brief description of the different extraction methods and detection techniques used for detecting insecticides in various environmental matrices.

Advances in the Determination of Insecticides Using Biosensors

Although chromatographic-based methods are traditional techniques popular for pesticide residue detection, the concentration of the samples is necessary to augment detection sensitivity, which in turn results in sample destruction many times. Additionally, the methods are relatively complicated, requiring expensive instruments, high volumes of toxic solvents for sample preparation, and trained personnel/technicians for instrument handling to ensure reliable analysis of pesticide residues. Therefore, efforts are being continuously made to create alternative systems that are fast, sensitive, precise, and easy to handle.

The advanced techniques that have been developed and are being adopted include biosensors. A biosensor is a technique that integrates biological entities with an electronic device for the rapid analysis of insecticides and their residues through signal detection. Biosensors can be classified into two classes based on the biorecognition element employed and the transducer used for signal amplification and detection. A brief description of each technique is mentioned below. Figure 3 illustrates the various types of biosensors, classified by biorecognition element and transducer.

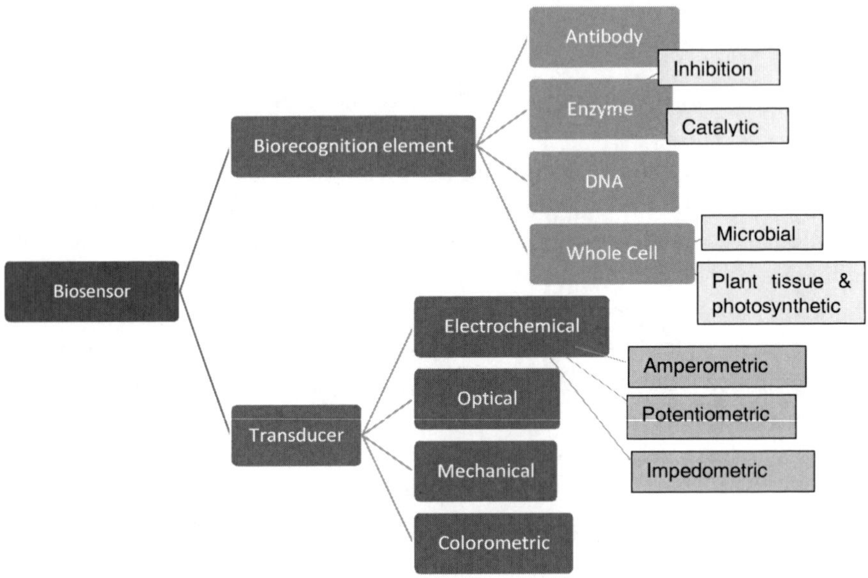

Figure 3. Classification of biosensors based on biorecognition element and transducer.

Biosensors Based on Biorecognition Elements

1. *Immunosensors:* An immunosensor is formed by the complex formation between antibodies incorporated on a substrate and the antigen or toxic compound to be detected. Due to the equilibrium association constant of the antibody, antigens can be easily detected. The signal developed due to the reaction between the antibody and the antigen can be detected and amplified using a suitable transducer, such as an electrochemical, optical, mechanical, or piezoelectric

transducer, for signal detection. Immunosensors are highly specific, with a low limit of detection and low cost, for determining the presence of a targeted compound. However, the generation of monoclonal antibodies is a limitation to the adoption of this technique on a larger scale. Also, for the detection of insecticides by immunosensors, which are smaller in size, an additional carrier protein molecule is required (Christopher et al., 2020).

2. *Enzymatic biosensors:* In this technique, enzymes are used as biomarkers to recognize and sense insecticide residues. Due to its unique properties of rapidity, high specificity, and sensitivity, this technique is the most popular method used for detecting insecticide residues in various matrices. The enzyme is attached to the surface of the electrode following the adsorption, entrapment, electro polymerization, or self-assembly of the monolayer. The effect of insecticide can be detected by two modes, *viz.* enzymatic activity inhibition (measurement of signal due to enzymatic activity is done in the presence as well as the absence of the inhibitor) and/or catalytic mode (in which the detection and quantification of the analyte/ substrate is done based on the enzyme being catalysed by it) (Amine et al., 2016). However, the sensitivity of the technique depends on the enzyme purity and other impurities, such as surfactants, additives, and heavy metals, that may be present along with the insecticides (Vohra et al., 2020).
3. *DNA-based sensors:* For sensing insecticides, the DNA molecule is immobilized on the working transducer surface. The change in the redox potential of the DNA is recorded using an electrochemical transducer with a potentiometric, voltammetric technique, or spectroscopic tagging of the nucleotide (Nowicka et al., 2010). These biosensors have been utilized for the rapid recognition of biological warfare agents, forensic sciences, and genetically modified foods (Lim et al., 2008; Stobiecka et al., 2007). However, due to the denaturation of DNA, the signal intensity gets reduced, thus limiting their application. In addition, these biosensors have issues with reusability and sensitivity, thus limiting their commercial application (He et al., 2005).
4. *Cell-based sensors:* Microorganisms, tissues from plants/animals, or living cells, either as a whole or in their subcellular fractions, are incorporated/ fixed on the surface of the electrode, either by crosslinking or entrapment, to detect insecticides or the target analyte.

Living cells/microorganisms are responsible for the oxidation of foreign compounds through an enzymatic reaction, the response of which is measured by recording the signal (Dai & Choi, 2013). In these sensors, there is no need to isolate/ purify the enzyme, as it remains in its natural environment, which improves its sensitivity and stability, making it economical and less laborious. The microbial sensors can also detect the toxicity as well as the bioavailability of the toxic compound being analysed. To enhance the sensing capability of the sensor, researchers have utilized genetically modified organisms (GMOs). Although microbial biosensors sideline the task of enzyme purification/ isolation, their low selectivity due to non-specific chemical reactions and slow response due to the diffusion of substrate and products through the cell wall limit their wide application (Audrey et al., 2012).

Transducer-Based Biosensors

1. Electrochemical biosensors are the most widely and commonly used biosensors for detecting insecticides and their residues, as they provide higher sensitivity, specificity, and reproducibility (Jain et al., 2022). The residue level is determined through electrochemical or immunoreactions, which are generated and measured by changes in electrical signals, such as potential, current, or impedance. In electrochemical biosensors, the enzyme activity has also been utilized to detect insecticide residue levels by combining electrochemical systems and enzymatic reactions. These sensors are further classified into three classes based on the signal properties. The classes include a) Amperometric sensors that measure the change in the current caused by the electrochemical reaction, b) Impedimetric biosensors that help to analyse the residues of the insecticides by measuring the changes in impedance and c) Potentiometric sensors that detect the residues by recording the changes in concentration and pH of the solution resulting from the reaction between the antibody and the antigen (Ba Hashwan et al., 2020).
2. Optical biosensors have been extensively utilized for the detection and monitoring of insecticide residues in various matrices. The technique detects the analyte based on changes in optical properties, such as absorption, emission, or reflectance, when the insecticide

interacts with biomolecules. For detecting chemical-physical changes due to insecticide interaction, optical biosensors utilize fluorescence absorption, SPR (surface plasmon resonance), SER (Surface-enhanced Raman Spectroscopy), and colorimetric techniques. A fluorescence technique was utilized by Paliwal et al. (2007) for the detection of p-nitrophenol and p-nitrophenol-substituent organophosphorus neurotoxins by quenching of coumarin1, which is a competitive inhibitor of organophosphorus hydrolase, an OP-hydrolyzing enzyme. A pH-sensitive fluorescent probe was utilized by Jin et al., (2004) to detect the inhibition of the enzyme acetylcholinesterase (AChE) by organophosphates and carbamates. The detection limits for carbofuran, carbaryl, paraoxon, and dichlorvos using the probe were found to be 3.5, 50, 12, and 25 μg/L in pure water, respectively.

The SPR technique involves capturing photons by free electrons and measuring the changes in resonance angle (θ) resulting from changes in the analyte concentration (Mello & Kubota, 2002). In contrast, the SER technique records changes in signal intensity. The colorimetric technique measures the color change that occurs as a result of the chemical reaction. The technique offers several advantages, including high sensitivity, specificity, selectivity, and real-time analysis of insecticides; however, specific materials with desirable electrical and optical properties are needed to develop these biosensors. Additionally, sophisticated instruments are required for characterizing the device.

3. *Mechanical/Mass-based/ Piezoelectric biosensors:* These biosensors record the minute changes on the probe or transducer surface used for fabricating the sensor. In this technique, the change in mechanical properties, such as mass, viscosity, or density, of the insecticide under study is measured. Piezoelectric sensors have the potential to detect insecticides and their residues in real-time, as they are sensitive, selective, and label-free sensors. Magnetoelastic biosensors have also been used for residue detection in different matrices. They have a thick film (magnetoelastic) as the key element of the biosensor, which vibrates mechanically and generates a characteristic frequency when an external magnetic field is applied after being coupled with the sensing film (Zeng & Grimes, 2004).

In recent times, microelectromechanical system (MEMS) resonant sensors have proven their capability as mass sensors (P. Kumar et al., 2015). In addition, micro-cantilever devices and thin-film piezoelectric MEMS resonant mass sensors have also found applications (Pang et al., 2012). However, the technique is limited by the challenges of imputing the actuation mode for the liquid environment, squeeze film damping, and electrolysis, which are major issues that need to be overcome.

Role of Nanotechnology in Biosensor Development

To improve detection and enhance the sensitivity and stability of biosensors, nanomaterials have been found to be effective in the fabrication of biosensors. The application of nanotechnology for their development enables the effective and enhanced recognition of insecticides due to their greater surface area, high electronic conductivity, and improved analytical performance (Justino et al., 2015). The utilization of nanomaterials on sensing platforms has led to the sensitive detection of target molecules through enhanced optical and electrochemical detection. Zhang (2013) developed a self-assembled monolayer of Ag nanoparticles as a Surface Enhanced Raman Scattering (SERS) substrate to detect methyl parathion. The results showed excellent Raman enhancement, with a limit of detection as low as 10^{-7} M. Zhao et al. (2015) fabricated a biosensor for the detection of organophosphates by direct electrodeposition of electrochemically reduced graphene oxide (ERGO)-Au nanoparticles (AuNPs)-β-cyclodextrin (β-CD) and Prussian blue-chitosan (PB-CS) on a glass carbon electrode (GCE) and fixing acetylcholinesterase (AChE) enzyme on the platform. The sensor showed wide linear ranges of $7.98–2.00\times10^{3}$ pg mL^{-1} and $4.3–1.00\times10^{3}$ pg mL^{-1} for malathion and carbaryl, respectively. The detection limits were also found to be low for both compounds. The biosensor was found to be highly sensitive and stable, with a rapid response time. Chang et al. (2017) developed a single-shot optical probe using fluorescent carbon dots for the rapid detection of paraoxon-ethyl in real vegetable samples. The limit of detection of the probe was found to be 0.22 ± 0.02 μM, whereas the dynamic linear range was found to be up to 5.80 mM.

Yan et al. (2017) developed a novel fluorimetric sensing platform for the quantitative detection of paraoxon through tyrosinase (TYR) enzyme-controlled quenching of gold nanoclusters (AuNCs). AuNCs were synthesised from chicken egg white (CEW) as a template and a stabilizer. The biosensor could detect paraoxon rapidly and possessed excellent sensitivity, having a detection limit of 0.1 ng mL^{-1}. Vaghela et al. (2018) developed a potentiometric biosensor based on agarose-guar gum (A-G) entrapped bio-nanoconjugate of urease enzyme with gold nanoparticles (AUNps) for the detection of glyphosate. The inhibitory urease activity by the analyte was measured by potentiometry through the use of an ammonium ion-selective electrode covered with an A-G-urease nanoconjugate membrane.

Table 3. Application of enzymatic nano-biosensors for the detection of insecticides in food commodities

Sensing electrode	Analyte	Commodity	Reference
AChE/WO_3/g-C_3N_4/GCN/PGE	Phosmet	Wheat grains	(Bilal et al., 2021; Raymundo-Pereira et al., 2021)
IL@$CoFe_2O_4$NPs@ MWCNTs@GCE	Fenitrothion	Orange, Grape	(Kilele et al., 2021)
ZnO-G/GCE	Paraoxon-Methyl	Cabbage, Tomato	(Parab et al., 2020)
Stearic acid/nanosilver/GCE	Paraoxon	Onion, Paddy grains	(Kumaravel et al., 2020)
ZIF-8/MB composites	Paraoxon	Apple, Eggplant	(X. Li et al., 2020)
SP-MCH-GE (DRAB + chlorpyrifos)	Chlorpyrifos	Apple, Orange, Cabbage	(W. Wang et al., 2020)
UiO-66-NH_2	Methyl parathion	Oranges, Tomatoes	(Mehta et al., 2019)
BSA/AChE-Glu-s-SWCNTs/GCE	Methyl parathion	Strawberry, Apple	(T. H. V. Kumar & Sundramoorthy, 2019)
AChE/ CS@TiO_2-CS/rGO/GCE	Dichlorvos	Cabbage juice	(Cui et al., 2018)
GC/rGO/AChE	Carbaryl	Tomato	(da Silva et al., 2018)
AChE/CNTs-NH_2/Ag NPs-N-F-MoS_2/GCE	Monocrotophos Chlorpyrifos	Apple	(Song et al., 2018)
AChE-e-pGON/GCE	Carbaryl	Cabbage, spinach	(Y. Li et al., 2017)
AChE/OMC-CS/Fe_3O_4-CS/SPCE	Methamidophos	Cabbage, Rape, Lettuce	(Q. Zhang et al., 2016)
AChE–MWCNTs/GONRs/GCE	Carbaryl	Cabbage	(Q. Liu et al., 2015)
CLDH-AChE/GN-AuNPs/GCE	Chlorpyrifos	Leek, Pakchoi	(Zhai et al., 2014)
AChE/PDDA-MWCNTs-GR/GCE	Carbaryl	Cabbage, Garland Chrysanthemum, Leek, Pakchoi	(X. Sun et al., 2013)
GCE/MWCNT/PANI/AChE	Carbaryl Methomyl	Apple, Broccoli, Cabbage	(Cesarino et al., 2012)

The developed sensor was found to have selectivity, reproducibility and prolonged stability of 180 days. Qin et al., (2020) developed a dissociable photoelectrode based on CdS nanocrystal-functionalized MnO_2 nanosheets for the detection of paraoxon that inhibited AChE activity and prevented the generation of TCh; thus, dissociation of MnO_2 nanosheets. Under the optimized parameters, the PEC sensor displayed a wide linear range of 0.05 to 10 ng/mL, and the detection limit was found to be 0.017 ng/mL. Also, the PEC sensor showed good sensitivity and stability that can be utilized for the analysis of real samples. Zhao et al. (2021) designed a nanogold/ mercaptomethamidophos multi-residue electrochemical biosensor that was able to simultaneously detect 11 kinds of organophosphorus pesticides (OPPs) and quantify the total amount of OPPs through an indirect competitive method. The biosensor could rapidly detect multiple pesticides and exhibited excellent electrochemical properties, having a linear range of 0.1 ~ 1500 ng/ mL and a lower detection limit of 0.019 ~ 0.077 ng/ mL with better stability and repeatability. The majority of the work employing nanomaterials has been utilized for the development of enzymatic sensors. Table 3 gives a brief description of different sensing electrodes and the utilization of enzymatic bio and nanobiosensors for the detection of various insecticides and their residues in different food commodities.

Innovations in the Field of Biosensors

a. Algae-based sensors

Algae-based biosensors are an attractive choice for pesticide detection as these are sustainable, possess high sensitivity, are cost-effective in terms of the cultivation of algae, and can be manipulated at the genetic level. In addition, they can adapt to diverse environments and be integrated with multiplexed devices and microfluidics to enhance their performance (Antonacci & Scognamiglio, 2020). Moro et al. (2018) used green photosynthetic microalgae-based optical bioassay for the rapid detection of marine pesticides. The pollutants could be detected within 10 min, and their concentrations ranged between ng/L to μg/L.

b. Nanozyme biosensors

Nanozymes are nanomaterials that possess enzyme-like properties and are also known as artificial enzymes. Nanozymes are gaining attention for biosensor development these days because they have high stability and sensitivity, are cost-effective, and easy to synthesise. These can also be combined with other bio-recognition materials to enhance the sensitivity of the current detection system (Zhao et al., 2023). Sun et al. (2021) prepared a bifunctional cerium oxide (CeO_2) nanozyme-modified glassy carbon electrode for the electrochemical detection of methyl-paraoxon. Cyclic voltammetry was applied to investigate the electrochemical behaviour of the modified electrode. Under the optimum conditions, the limit of detection of the sensor was found to be 0.06 mmol/ L with a linear range of 0.1 to 100 mmol/ L. Chen et al. (2021) developed a multiplex fluorescent bio-barcode immunoassay for the simultaneous detection of parathion, triazophos, and chlorpyrifos using AuNP signal amplification technology and Au@Pt catalysis. The recoveries of the developed assay ranged from 71.26 to 117.47%, with RSD values of 2.52 to 14.52%. The limit of detection of parathion, triazophos, and chlorpyrifos was found to be 9.88, 3.91, and 1.47 ng/kg, respectively.

c. Aptamer-based sensors

Aptamers are single-stranded oligonucleotides that possess high specificity and affinity for the target analytes. These are low-cost materials having satisfactory reproducibility and short detection time, thus making them useful as potential recognition elements in biosensors to be used for the detection of contaminants like insecticides in different environmental components. Aptamer binding to the contaminant results in conformational changes that are detected through the transducer. They can attain particular molecule detection potential by chemical bonding, which may be H-bonding, π-π bonding, van der Waals forces, or hydrophobic interactions (Kadam & Hong, 2022). Wang et al. (2019) developed a dual-modal sensor combined with ssDNA aptamer/AuNPs-based colorimetry and PLNRs-based phosphorescence inner filter effect(PIFE) for the detection of isocarbophos in vegetables. The assay was found to show good linearity in the range of 50–500 μg/L in colorimetry and 5–160 μg/L in phosphorescence mode, respectively. The limit of detection of 7.1 μg/L and 0.54 μg/L was found for the two modes. Xu et al. (2020) developed an electrochemical aptasensor by immobilizing acetamiprid-binding on highly porous gold through self-assembly to determine acetamiprid in fruits and vegetables. Using optimized parameters, the aptasensor was found to show a linear concentration range of 0.5–300 nmol/ L, whereas the detection limit was found to be 0.34 nmol/L. The aptasensor was found to have high selectivity with good reproducibility and stability.

d. Molecularly Imprinted Polymer (MIP) based sensors

Molecularly imprinted polymers (MIPs) are synthetic polymers developed through the molecular imprinting technique of the target template/ analyte to mimic the bioreceptor so as to detect, enrich, and separate the compound of interest (Fang et al., 2021). MIP-based sensors have been utilized for monitoring environmental contaminants, including pesticides and insecticides, as these sensors are quick, simple, stable, and cost-effective. Sroysee et al., (2019) developed molecularly imprinted polymers (MIP) separately for the selective binding of the insecticides carbofuran (CBF) and profenofos (PFF) by spin-coating them on one electrode pair of a dual-electrode QCM. The resulting pesticide QCM sensors showed a dynamic detection range of 0.5–1000 μM for CBF-MIP and 10–1000 μM for PFF-MIP. It was also found that both the sensors exhibited signals that were ten times higher than their non-imprinted counterpart. Li et al., (2019) prepared a reagentless electrochemiluminescence sensor with molecular imprinting, an electropolymerized poly (luminol-*p*-aminothiophenol) composite-modified gold electrode for the detection of triazophos in water samples. The results showed that the imprinted poly(luminol-*p*-aminothiophenol) displayed better ECL emission to triazophos as compared to polyluminol. The sensor showed a linear range from 1.0×10^{-10} to 1.0×10^{-6} M with a lower detection limit of 5.8×10^{-11} M for triazophos.

e. Metal-organic framework (MOF) based sensors

MOF-based sensors are porous materials which are having crystalline frameworks and are formed by the combination of metal ions/ clusters with organic ligands *via* coordination bonds. These are self-assembled materials that possess high specific area and catalytic activity along with high porosity, tunable polarity, and stability (thermal as well as chemical) (Haider et al., 2022). As MOFs possess high crystallinity and permanent porosity, these materials can analyze different target compounds by detecting the signals produced due to changes in their optical, mechanical, catalytic or conductive properties (W. Wu et al., 2023). Cao et al., (2022) developed a colourimetric detection platform based on a multi-enzyme cascade for acephate detection by encapsulating three enzymes, *viz.,* acetylcholinesterase, horseradish peroxidase, and choline oxidase, in a cruciate flower-like zeolitic imidazolate framework-8 (CF-ZIF-8) with acetylcholine chloride as substrate. The results showed that the acephate inhibited the activity of acetylcholinesterase, obstructed the cascade reaction, and reduced the production of H_2O_2, resulting in the intensity of colour, which was detected through colourimetry. The multi-enzyme platform was found to have a detection limit of 0.23nM for acephate and was used for pesticide detection in Chinese cabbage and romaine. (Hou et al., 2023) developed a colourimetric method for the visual detection of profenofos on Zn-ZrMOF containing Lewis acid-base catalytic sites for hydrolysis. The method was found to have a linear range of 0.01–1000 μg/ mL, whereas the detection limit was found to be 6.33 × 10−3 μg/ mL. In addition, the method showed good selectivity and accuracy for the determination of profenofos.

Table 4 briefly encompasses the biosensors being developed and utilized for the different insecticides in various food and environmental matrices. All these techniques have aided in the development of new methodologies for the rapid detection and analysis of insecticides and their residues in different matrices efficiently.

Role of Big Data Analysis and Artificial Intelligence in Residue Detection

The emergence and development of new detection techniques, in addition to available techniques for insecticides and their residue detection in different commodities and matrices, as discussed above, have led to enhanced accuracy. However, it has also led to the generation of huge amounts of analytical data. The analytical data are intertwined with the practices of pesticide usage, impact on crop production, insect-pest population, and geographic and climatic conditions, resulting in the generation of large and complex datasets (Y. Chen et al., 2020). To disseminate the information and make it available and useful to researchers working in different parts of the world, it becomes imperative to process, analyse, store, and manage the big data efficiently (Kote, 2018). Integration of statistically analysed data with data mining can lead to the development of smart information systems that can help to improve data processing. Quality data management will help in the utilization of data resources and offer support for decision-making for improvement in agricultural practices at the grassroots level, and open the way for adopting smart and sustainable agriculture (Jararweh et al., 2023). Smart agriculture is defined as the management approach that will empower the agriculture sector by automatically tracking, monitoring, and controlling logically and wisely to make decisions with the support of information and knowledge (Alrowais et al., 2022). In this direction, researchers have taken the initiative by developing PRIAS, an intelligent analysis system for pesticide residue detection data (Y. Chen et al., 2022). It has the ability for online collection, fusion processing, storage, and intelligent analysis of pesticide residue detection data. It not only carries out the comprehensive analysis from multiple outlooks but also explores the implicit associations in the data with the generation of analysis reports automatically.

Bottleneck and Future Perspectives in Residue Detection and Analysis

Despite tremendous advances in the development of on-site detection methods, there is still a need to address the limitations of current approaches, including handling and pre-treatment of the samples, extraction efficiency, consistency, reproducibility, and cost-effectiveness. Also, the methods developed so far have shown good results in the laboratory, but their efficiency still needs to be assessed in field conditions. In addition, nanomaterials are being employed by various researchers for the development of recognition elements or biosensors to detect different environmental contaminants in various environmental matrices and commodities, and the results have also been found to be encouraging. However, the reusability efficiency, precision, and accuracy in the detection, cost-effectiveness, disposal, and safety concerns associated with the nanomaterials need to be explored and addressed efficiently. The researchers need to look into a holistic approach for the development of materials for the detection and analysis of safety concerns as well.

Though initiatives have been taken in the development of safe, green, and sustainable technologies for insecticide detection, it needs to be addressed on a larger scale. In addition, the huge data generated needs to be compiled in conjunction with the Internet of Things (IoT) and transferred to researchers involved in this field so as to utilise the data efficiently for the development of environment-friendly techniques and adopt smart agriculture for the betterment of the farming community as well as consumers.

Table 4. Different biosensors used for insecticide detection in different environmental matrices

Immunosensors			
Type	Insecticide/ pesticide detected	Method of transduction/detection	Reference
Electrochemical	2,4-Dichlorophenoxy acetic acid	Amperometric	(Kröger et al., 1998)
	Atrazine		(Grennan et al., 2003)
	Atrazine	Conductometric	(Valera et al., 2010)
	Terbuthylazine (TBA)	Potentiometric	(Mosiello et al., 2003)
	Atrazine 2,4-Dichlorophenoxy acetic acid	Electrochemical impedance spectroscopy	(Navrátilová & Skládal, 2004)
Optical	DTT, Chlorpyrifos, and Carbaryl	Surface plasmon resonance	(Ramón-Azcón et al., 2008)
	Atrazine	Fluorescence polarisation	(Cummins et al., 2006)
	Atrazine, simazine, and alachlor	Total internal reflection fluorescence (TIRF)	(Mallat et al., 2001))
	Atrazine	Polarisation-modulation infrared reflection-absorption spectroscopy (PM-IRRA)	(Salmain et al., 2008)
Piezoelectric immunosensors	Chlorpyrifos, Triclopyr		(J. Kaur et al., 2004)
Mechanical Immunosensors	Atrazine		(Raman Suri et al., 2008)
	DTT		(Alvarez et al., 2003)
	Atrazine, 2,4-Dichlorophenoxy acetic acid		(J. Kaur et al., 2004)
Nanoparticles based biosensors			
Gold nanoparticles (AuNP)	Herbicide simazine	Electrochemical	(J. Zhang et al., 2017)
	Organophosphate pesticides	Colorimetric	(S. Wu et al., 2017)
	Methyl parathion	Electrochemical	(G. Liu et al., 2014)
	Methyl paraoxon Carbofuran Phoxim	Amperometric	(Yin et al., 2009)
	Dichlorodiphenyltrichloroethane (DDT)	Dipstick immunoassay	(Lisa et al., 2009)
	Paraoxon	Electrochemical	(Hu et al., 2003)

Table 4. (Continued)

Type	Insecticide/ pesticide detected	Method of transduction/detection	Reference
	Paraoxon	Amperometric	(Albareda-Sirvent et al., 2001)
	Carbofuran		
Gold nanoparticles/dragonfly arrays	Rhodamine	Surface-Enhanced Raman Spectroscopy (SERS)	(Wei et al., 2016)
Fe_3O_4 functionalized graphene oxide – AuNP.	Catechol Hydroquinone	Electrochemical	(Erogul et al., 2015)
4-amini-3-mercaptobenzoic acid functionalized AuNP	Cyhalothrin	Colorimetric	(Imene et al., 2014)
Au-Na dodecylbenzene sulphonate nanoparticles	Methyl parathion	Electrochemical	(C. Li et al., 2011)
CdTe quantum dots/ AuNPs	Monocrotrophos	Amperometric	(Du et al., 2008)
Chitosan-TiO_2- graphene nanocomposites	Dichlorvos	Amperometric	(Cui et al., 2018)
Chitosan-TiO_2-gold nanorods@ mesoporous SiO_2	Fenthion Dichlorvos	Amperometric	(Cui et al., 2018)
Gold nanoparticle	Malathion Methyl parathion	Amperometric	(Jiang et al., 2018)
Gold nanoparticles, Multi-walled Carbon Nanotubes on a gold wire	Methyl Parathion Monocrotophos Chlorpyrifos Endosulfan	Amperometric	(Dhull, 2020)
Antimony tin oxide chitosan (ATO-CS) and mesoporous carbon-chitosan (OMC-CS) composite)	Chlorpyrifos Methamidophos	Amperometric	(Hou et al., 2020)
Graphene/copper oxide nano-flowers	Malathion	Amperometric 3D	(Bao et al., 2019)
Prussian blue nanoparticles, Carbon black	2,4-D Atrazine Paraoxon	Amperometric	(Arduini et al., 2019)
Graphene, Platinum	Paraoxon	Amperometric	(Hondred et al., 2018)
Urease with gold	Glyphosate	Potentiometric	(Vaghela et al., 2018)
MWCNTs and poly-Lysine	Diazinon	Impedimetric	(Zare et al., 2019)
Gold	Chlorpyrifos	Impedimetric	(Hou et al., 2020)

Type	Insecticide/ pesticide detected	Method of transduction/detection	Reference
Graphene oxide, copper	Profenofos Phorate Isocarbophos Omethoate	Impedimetric	(Fu et al., 2019)
MIP-based biosensors			
Silica nanoparticles	Pyrethroid (3-phenoxybenzaldehyde)	Colorimetric	(Ye et al., 2018)
SiO_2@QDs@m-MIPs	2,4-dichlorophenoxyaceticacid	Fluorescence	(Jia et al., 2017)
Methacrylic acid-based m-MIP	Methyl parathion	Electrochemical	(Hassan et al., 2018)
Photonic hydrogel film	Imidacloprid	Optical	(X. Wang et al., 2013)
AuNPs/ERGO-SPCE*	Cyhexatin	Electrochemical	(C. Zhang et al., 2019)
MIP coated-QDs	Cyphenothran	Fluorescence	(Ren & Chen, 2015)
Ag-N@ZnO/CHAC#	Cypermethrin	Electrochemical	Zhang et al., (2019)
MWCNT-MIP	Lindane	Potentiometric	(Anirudhan & Alexander, 2015)
MWCNT-MIP	Dicloran	Voltammetric	(Shahtaheri et al., 2017)
Molecularly imprinted film	Methyl parathion	Optical	(Y. Tan et al., 2015)
Aptamers-based biosensors			
Platinum	Acetamiprid	Impedimetric	(Madianos et al., 2018)
	Atrazine		
Microcantilever array sensor	Profenofos	Optical	(C. Li et al., 2018)
Silver	Malathion	Surface-enhanced Raman scattering	(Nie et al., 2018)
Gold	Malathion	Colorimetric	(Bala et al., 2017)
Platinum	Acetamiprid	Impedimetric	(Nie et al., 2018)
	Atrazine		
GO-CuNPs (Graphene oxide-copper nanoparticles)	Prophenofos	Co-electrodeposition	(Fu et al., 2019)
	Phorate		
	Isocarbophos		
	Omethoate		

References

Albareda-Sirvent, M., Merkoçi, A., & Alegret, S. (2001). Pesticide determination in tap water and juice samples using disposable amperometric biosensors made using thick-film technology. *Analytica Chimica Acta*, *442*(1), 35–44. https://doi.org/10.1016/S0003-2670(01)01017-0.

Alrowais, F., Asiri, M. M., Alabdan, R., Marzouk, R., Hilal, A. M., Alkhayyat, A., & Gupta, D. (2022). Hybrid leader based optimization with deep learning driven weed detection on internet of things enabled smart agriculture environment. *Computers and Electrical Engineering*, *104*, 108411. https://doi.org/10.1016/j.compeleceng.2022.108411.

Alvarez, M., Calle, A., Tamayo, J., Lechuga, L. M., Abad, A., & Montoya, A. (2003). Development of nanomechanical biosensors for detection of the pesticide DDT. *Biosensors and Bioelectronics*, *18*(5–6), 649–653. https://doi.org/10.1016/S0956-5663(03)00035-6.

Amine, A., Arduini, F., Moscone, D., & Palleschi, G. (2016). Recent advances in biosensors based on enzyme inhibition. *Biosensors and Bioelectronics*, *76*, 180–194. https://doi.org/10.1016/j.bios.2015.07.010.

Anirudhan, T. S., & Alexander, S. (2015). Design and fabrication of molecularly imprinted polymer-based potentiometric sensor from the surface modified multiwalled carbon nanotube for the determination of lindane (γ-hexachlorocyclohexane), an organochlorine pesticide. *Biosensors and Bioelectronics*, *64*, 586–593. https://doi.org/10.1016/j.bios.2014.09.074.

Antonacci, A., & Scognamiglio, V. (2020). Biotechnological Advances in the Design of Algae-Based Biosensors. *Trends in Biotechnology*, *38*(3), 334–347. https://doi.org/10.1016/j.tibtech.2019.10.005.

Arduini, F., Cinti, S., Caratelli, V., Amendola, L., Palleschi, G., & Moscone, D. (2019). Origami multiple paper-based electrochemical biosensors for pesticide detection. *Biosensors and Bioelectronics*, *126*, 346–354. https://doi.org/10.1016/j.bios.2018.10.014.

Audrey, S., Beatriz, P.-S., & Jean-Louis, M. (2012). Biosensors for Pesticide Detection: New Trends. *American Journal of Analytical Chemistry*, *2012*(03), 210–232. https://doi.org/10.4236/AJAC.2012.33030.

Aysal, P., Arp´ad, ´, Ambrus, A. A., Lehotay, S. J., & Cannavan, A. (2007). Validation of an efficient method for the determination of pesticide residues in fruits and vegetables using ethyl acetate for extraction. *Taylor & FrancisP Aysal, Á Ambrus, SJ Lehotay, A CannavanJournal of Environmental Science and Health, Part B, 2007•Taylor & Francis*, *42*(5), 481–490. https://doi.org/10.1080/19312450701392490.

Ba Hashwan, S. S., Khir, M. H. B. M., Al-Douri, Y., & Ahmed, A. Y. (2020). Recent progress in the development of biosensors for chemicals and pesticides detection. *IEEE Access*, *8*, 82514–82527. https://doi.org/10.1109/ACCESS.2020.2991380.

Bahaghighat, H. D., Freye, C. E., & Synovec, R. E. (2019). Recent advances in modulator technology for comprehensive two dimensional gas chromatography. *TrAC Trends in Analytical Chemistry*, *113*, 379–391. https://doi.org/10.1016/J.TRAC.2018.04.016.

Bala, R., Dhingra, S., Kumar, M., Bansal, K., Mittal, S., Sharma, R. K., & Wangoo, N. (2017). Detection of organophosphorus pesticide – Malathion in environmental samples using peptide and aptamer based nanoprobes. *Chemical Engineering Journal, 311*, 111–116. https://doi.org/10.1016/j.cej.2016.11.070.

Bala, R., Mittal, S., Sharma, R. K., & Wangoo, N. (2018). A supersensitive silver nanoprobe based aptasensor for low cost detection of malathion residues in water and food samples. *Spectrochimica Acta - Part A: Molecular and Biomolecular Spectroscopy, 196*, 268–273. https://doi.org/10.1016/j.saa.2018.02.007.

Bao, J., Huang, T., Wang, Z., Yang, H., Geng, X., Xu, G., Samalo, M., Sakinati, M., Huo, D., & Hou, C. (2019). 3D graphene/copper oxide nano-flowers based acetylcholinesterase biosensor for sensitive detection of organophosphate pesticides. *Sensors and Actuators, B: Chemical, 279*, 95–101. https://doi.org/10.1016/j.snb.2018.09.118.

Bernhardt, E. S., Rosi, E. J., & Gessner, M. O. (2017). Synthetic chemicals as agents of global change. *Frontiers in Ecology and the Environment, 15*(2), 84–90. https://doi.org/10.1002/FEE.1450.

Bilal, S., Mudassir Hassan, M., Fayyaz ur Rehman, M., Nasir, M., Jamil Sami, A., & Hayat, A. (2021). An insect acetylcholinesterase biosensor utilizing WO3/g-C3N4 nanocomposite modified pencil graphite electrode for phosmet detection in stored grains. *Food Chemistry, 346*. https://doi.org/10.1016/j.foodchem.2020.128894.

Bilehal, D. C., Chetti, M. B., Sung, D. D., & Goroji, P. T. (2014). Reversed-phase UPLC method for the determination of monocrotophos, thiram, carbendazim, carbaryl, and imidacloprid pesticides in mango and pomegranate by QuEChERS method. *Journal of Liquid Chromatography and Related Technologies, 37*(12), 1633–1643. https://doi.org/10.1080/10826076.2013.803205.

Busa, L. S. A., Mohammadi, S., Maeki, M., Ishida, A., Tani, H., & Tokeshi, M. (2016). Advances in microfluidic paper-based analytical devices for food and water analysis. *Micromachines, 7*(5). https://doi.org/10.3390/mi7050086.

C, S. R., & G V, S. R. (2017). Pesticide Residues Recovery studies by QuEChERS Sample Technique Using Gas Chromatography. *Madras Agricultural Journal, 104*(.1-.4). https://doi.org/10.29321/maj.01.000400.

Cao, X., Guo, Y., Zhao, M., Li, J., Wang, C., Xia, J., Zou, T., & Wang, Z. (2022). An efficient multi-enzyme cascade platform based on mesoporous metal-organic frameworks for the detection of organophosphorus and glucose. *Food Chemistry, 381*(January). https://doi.org/10.1016/j.foodchem.2022.132282.

Cervera, M. I., Portolés, T., Pitarch, E., Beltrán, J., & Hernández, F. (2012). Application of gas chromatography time-of-flight mass spectrometry for target and non-target analysis of pesticide residues in fruits and vegetables. *Journal of Chromatography A, 1244*, 168–177. https://doi.org/10.1016/j.chroma.2012.04.063.

Cesarino, I., Moraes, F. C., Lanza, M. R. V., & MacHado, S. A. S. (2012). Electrochemical detection of carbamate pesticides in fruit and vegetables with a biosensor based on acetylcholinesterase immobilised on a composite of polyaniline-carbon nanotubes. *Food Chemistry, 135*(3), 873–879. https://doi.org/10.1016/j.foodchem.2012.04.147.

Chang, M. M. F., Ginjom, I. R., & Ng, S. M. (2017). Single-shot 'turn-off' optical probe for rapid detection of paraoxon-ethyl pesticide on vegetable utilising fluorescence carbon dots. *Sensors and Actuators, B: Chemical, 242*, 1050–1056. https://doi.org/10.1016/j.snb.2016.09.147.

Chen, G., Liu, G., Jia, H., Cui, X., Wang, Y., Li, D., Zheng, W., She, Y., Xu, D., Huang, X., Abd El-Aty, A. M., Sun, J., Liu, H., Zou, Y., Wang, J., Jin, M., & Hammock, B. D. (2021). A sensitive bio-barcode immunoassay based on bimetallic Au@Pt nanozyme for detection of organophosphate pesticides in various agro-products. *Food Chemistry*, *362*(May), 130118. https://doi.org/10.1016/j.foodchem.2021.130118.

Chen, Y., Dou, H., Chang, Q., & Fan, C. (2022). PRIAS: An Intelligent Analysis System for Pesticide Residue Detection Data and Its Application in Food Safety Supervision. *Foods*, *11*(6), 1–15. https://doi.org/10.3390/foods11060780.

Chen, Y., Lv, C., Li, Y., Chen, W., & Ma, K. L. (2020). Ordered matrix representation supporting the visual analysis of associated data. *Science China Information Sciences*, *63*(8). https://doi.org/10.1007/s11432-019-2647-3.

Christopher, F. C., Kumar, P. S., Christopher, F. J., Joshiba, G. J., & Madhesh, P. (2020). Recent advancements in rapid analysis of pesticides using nano biosensors: A present and future perspective. *Journal of Cleaner Production*, *269*, 122356. https://doi.org/10.1016/j.jclepro.2020.122356.

Chui, M. Q., Thang, L. Y., & See, H. H. (2017). Integration of the free liquid membrane into electrokinetic supercharging – capillary electrophoresis for the determination of cationic herbicides in environmental water samples. *Journal of Chromatography A*, *1481*, 145–151. https://doi.org/10.1016/j.chroma.2016.12.042.

Cui, H. F., Wu, W. W., Li, M. M., Song, X., Lv, Y., & Zhang, T. T. (2018). A highly stable acetylcholinesterase biosensor based on chitosan-TiO2-graphene nanocomposites for detection of organophosphate pesticides. *Biosensors and Bioelectronics*, *99*, 223–229. https://doi.org/10.1016/j.bios.2017.07.068.

Cummins, C. M., Koivunen, M. E., Stephanian, A., Gee, S. J., Hammock, B. D., & Kennedy, I. M. (2006). Application of europium(III) chelate-dyed nanoparticle labels in a competitive atrazine fluoroimmunoassay on an ITO waveguide. *Biosensors and Bioelectronics*, *21*(7), 1077–1085. https://doi.org/10.1016/j.bios.2005.04.003.

da Silva, M. K. L., Vanzela, H. C., Defavari, L. M., & Cesarino, I. (2018). Determination of carbamate pesticide in food using a biosensor based on reduced graphene oxide and acetylcholinesterase enzyme. *Sensors and Actuators, B: Chemical*, *277*, 555–561. https://doi.org/10.1016/j.snb.2018.09.051.

Dai, C., & Choi, S. (2013). Technology and Applications of Microbial Biosensor. *Open Journal of Applied Biosensor*, *2013*(03), 83–93. https://doi.org/10.4236/OJAB.2013.23011.

de Bon, H., Huat, J., Parrot, L., Sinzogan, A., Martin, T., Malézieux, E., & Vayssières, J. F. (2014). Pesticide risks from fruit and vegetable pest management by small farmers in sub-Saharan Africa. A review. *Agronomy for Sustainable Development*, *34*(4), 723–736. https://doi.org/10.1007/S13593-014-0216-7/FIGURES/4.

de Oliveira, M. L. G., Madureira, F. D., Aurélio, F., Pontelo, A. P., Silva, G., Oliveira, R., & Paes, C. (2012). A multi-residue method for the determination of pesticides in high water content matrices by gas chromatography-single quadrupole mass spectrometry with electron ionisation (EI-GC/MS). *Food Additives and Contaminants - Part A*, *29*(4), 657–664. https://doi.org/10.1080/19440049.2011.642102.

Dhull, V. (2020). A Nafion/AChE-cSWCNT/MWCNT/Au-based amperometric biosensor for the determination of organophosphorous compounds. *Environmental Technology (United Kingdom)*, *41*(5), 566–576. https://doi.org/10.1080/09593330.2018.1505964.

Du, D., Chen, S., Song, D., Li, H., & Chen, X. (2008). Development of acetylcholinesterase biosensor based on CdTe quantum dots/gold nanoparticles modified chitosan microspheres interface. *Biosensors and Bioelectronics*, *24*(3), 475–479. https://doi.org/10.1016/j.bios.2008.05.005.

Dusfour, I., Vontas, J., David, J. P., Weetman, D., Fonseca, D. M., Corbel, V., Raghavendra, K., Coulibaly, M. B., Martins, A. J., Kasai, S., & Chandre, F. (2019). Management of insecticide resistance in the major Aedes vectors of arboviruses: Advances and challenges. *PLOS Neglected Tropical Diseases*, *13*(10), e0007615. https://doi.org/10.1371/JOURNAL.PNTD.0007615.

Eissa, S., & Zourob, M. (2017). Selection and Characterization of DNA Aptamers for Electrochemical Biosensing of Carbendazim. *Analytical Chemistry*, *89*(5), 3138–3145. https://doi.org/10.1021/acs.analchem.6b04914.

Erogul, S., Bas, S. Z., Ozmen, M., & Yildiz, S. (2015). A new electrochemical sensor based on Fe3O4 functionalized graphene oxide-gold nanoparticle composite film for simultaneous determination of catechol and hydroquinone. *Electrochimica Acta*, *186*, 302–313. https://doi.org/10.1016/j.electacta.2015.10.174.

Fang, L., Jia, M., Zhao, H., Kang, L., Shi, L., Zhou, L., & Kong, W. (2021). Molecularly imprinted polymer-based optical sensors for pesticides in foods: Recent advances and future trends. *Trends in Food Science and Technology*, *116*(July), 387–404. https://doi.org/10.1016/j.tifs.2021.07.039.

Farajzadeh, M. A., Safi, R., & Yadeghari, A. (2019). Combination of QuEChERS extraction with magnetic solid phase extraction followed by dispersive liquid–liquid microextraction as an efficient procedure for the extraction of pesticides from vegetable, fruit, and nectar samples having high content of solids. *Microchemical Journal*, *147*, 571–581. https://doi.org/10.1016/j.microc.2019.03.074.

Fu, J., An, X., Yao, Y., Guo, Y., & Sun, X. (2019). Electrochemical aptasensor based on one step co-electrodeposition of aptamer and GO-CuNPs nanocomposite for organophosphorus pesticide detection. *Sensors and Actuators, B: Chemical*, *287*, 503–509. https://doi.org/10.1016/j.snb.2019.02.057.

Grennan, K., Strachan, G., Porter, A. J., Killard, A. J., & Smyth, M. R. (2003). Atrazine analysis using an amperometric immunosensor based on single-chain antibody fragments and regeneration-free multi-calibrant measurement. *Analytica Chimica Acta*, *500*(1–2), 287–298. https://doi.org/10.1016/S0003-2670(03)00942-5.

Guo, X., Wang, K., Chen, G. H., Shi, J., Wu, X., Di, L. L., & Wang, Y. (2017a). Determination of strobilurin fungicide residues in fruits and vegetables by nonaqueous micellar electrokinetic capillary chromatography with indirect laser-induced fluorescence. *Electrophoresis*, *38*(16), 2004–2010. https://doi.org/10.1002/elps.201700060.

Guo, X., Wang, K., Chen, G. H., Shi, J., Wu, X., Di, L. L., & Wang, Y. (2017b). Determination of strobilurin fungicide residues in fruits and vegetables by nonaqueous micellar electrokinetic capillary chromatography with indirect laser-induced fluorescence. *Electrophoresis, 38*(16), 2004–2010. https://doi.org/10.1002/elps.201700060.

Haider, J., Shahzadi, A., Akbar, M. U., Hafeez, I., Shahzadi, I., Khalid, A., Ashfaq, A., Ahmad, S. O. A., Dilpazir, S., Imran, M., Ikram, M., Ali, G., Khan, M., Khan, Q., & Maqbool, M. (2022). A review of synthesis, fabrication, and emerging biomedical applications of metal-organic frameworks. In *Biomaterials Advances* (Vol. 140). https://doi.org/10.1016/j.bioadv.2022.213049.

Hancu, G., Simon, B., Rusu, A., Mircia, E., & Gyéresi, Á. (2013). Principles of micellar electrokinetic capillary chromatography applied in pharmaceutical analysis. *Advanced Pharmaceutical Bulletin, 3*(1), 1–8. https://doi.org/10.5681/apb.2013.001.

Hassan, A. H. A., Moura, S. L., Ali, F. H. M., Moselhy, W. A., Taboada Sotomayor, M. del P., & Pividori, M. I. (2018). Electrochemical sensing of methyl parathion on magnetic molecularly imprinted polymer. *Biosensors and Bioelectronics, 118*, 181–187. https://doi.org/10.1016/j.bios.2018.06.052.

He, P., Xu, Y., & Fang, Y. (2005). A review: Electrochemical DNA biosensors for sequence recognition. In *Analytical Letters* (Vol. 38, Issue 15, pp. 2597–2623). https://doi.org/10.1080/00032710500369794.

Hondred, J. A., Breger, J. C., Alves, N. J., Trammell, S. A., Walper, S. A., Medintz, I. L., & Claussen, J. C. (2018). Printed Graphene Electrochemical Biosensors Fabricated by Inkjet Maskless Lithography for Rapid and Sensitive Detection of Organophosphates. *ACS Applied Materials and Interfaces, 10*(13), 11125–11134. https://doi.org/10.1021/acsami.7b19763.

Hou, L., Chen, X., Zhang, X., Tang, Y., Yao, Y., Lin, T., & Zhao, S. (2023). Zn-ZrMOF containing Lewis acid-base catalytic sites for hydrolysis and visual detection of profenofos. *Sensors and Actuators B: Chemical, 375*. https://doi.org/10.1016/j.snb.2022.132902.

Hou, L., Zhang, X., Kong, M., Jiang, G., Sun, Y., Mo, W., Lin, T., Ye, F., & Zhao, S. (2020). A competitive immunoassay for electrochemical impedimetric determination of chlorpyrifos using a nanogold-modified glassy carbon electrode based on enzymatic biocatalytic precipitation. *Microchimica Acta, 187*(4). https://doi.org/10.1007/s00604-020-4175-1.

Hu, S. Q., Xie, J. W., Xu, Q. H., Rong, K. T., Shen, G. L., & Yu, R. Q. (2003). A label-free electrochemical immunosensor based on gold nanoparticles for detection of paraoxon. *Talanta, 61*(6), 769–777. https://doi.org/10.1016/S0039-9140(03)00368-0.

Imene, B., Cui, Z., Zhang, X., Gan, B., Yin, Y., Tian, Y., Deng, H., & Li, H. (2014). 4-Amino-3-mercaptobenzoic acid functionalized gold nanoparticles: Synthesis, selective recognition and colorimetric detection of cyhalothrin. In *Sensors and Actuators, B: Chemical* (Vol. 199, pp. 161–167). https://doi.org/10.1016/j.snb.2014.03.097.

Jain, U., Saxena, K., Hooda, V., Balayan, S., Singh, A. P., Tikadar, M., & Chauhan, N. (2022). Emerging vistas on pesticides detection based on electrochemical biosensors – An update. *Food Chemistry, 371*(September 2021), 131126. https://doi.org/10.1016/j.foodchem.2021.131126.

Jararweh, Y., Fatima, S., Jarrah, M., & AlZu'bi, S. (2023). Smart and sustainable agriculture: Fundamentals, enabling technologies, and future directions. *Computers and Electrical Engineering, 110*, 108799. https://doi.org/10.1016/j.compeleceng.2023.108799.

Jia, M., Zhang, Z., Li, J., Shao, H., Chen, L., B, X. Y.-S. and A., & 2017, undefined. (2017). A molecular imprinting fluorescence sensor based on quantum dots and a mesoporous structure for selective and sensitive detection of 2, 4-dichlorophenoxyacetic. *Elsevier, 252*, 934–943. https://doi.org/10.1016/j.snb.2017.06.090.

Jiang, B., Dong, P., & Zheng, J. (2018). A novel amperometric biosensor based on covalently attached multilayer assemblies of gold nanoparticles, diazo-resins and acetylcholinesterase for the detection of organophosphorus pesticides. *Talanta, 183*, 114–121. https://doi.org/10.1016/j.talanta.2018.02.016.

Jin, S., Xu, Z., Chen, J., Liang, X., Wu, Y., & Qian, X. (2004). Determination of organophosphate and carbamate pesticides based on enzyme inhibition using a pH-sensitive fluorescence probe. *Analytica Chimica Acta, 523*(1), 117–123. https://doi.org/10.1016/j.aca.2004.05.030.

Justino, C. I. L., Freitas, A. C., Pereira, R., Duarte, A. C., & Rocha Santos, T. A. P. (2015). Recent developments in recognition elements for chemical sensors and biosensors. In *TrAC - Trends in Analytical Chemistry* (Vol. 68, pp. 2–17). Elsevier B.V. https://doi.org/10.1016/j.trac.2015.03.006.

Kadam, U. S., & Hong, J. C. (2022). Advances in aptameric biosensors designed to detect toxic contaminants from food, water, human fluids, and the environment. In *Trends in Environmental Analytical Chemistry* (Vol. 36, p. e00184). Elsevier. https://doi.org/10.1016/j.teac.2022.e00184.

Kaur, J., Singh, K. V., Schmid, A. H., Varshney, G. C., Suri, C. R., & Raje, M. (2004). Atomic force spectroscopy-based study of antibody pesticide interactions for characterization of immunosensor surface. *Biosensors and Bioelectronics, 20*(2), 284–293. https://doi.org/10.1016/j.bios.2004.01.012.

Kaur, R., Mavi, G. K., Raghav, S., & Khan, I. (2019). Pesticides Classification and its Impact on Environment. *International Journal of Current Microbiology and Applied Sciences, 8*(03), 1889–1897. https://doi.org/10.20546/ijcmas.2019.803.224.

Kilele, J. C., Chokkareddy, R., & Redhi, G. G. (2021). Ultra-sensitive electrochemical sensor for fenitrothion pesticide residues in fruit samples using IL@CoFe2O4NPs@MWCNTs nanocomposite. *Microchemical Journal, 164*. https://doi.org/10.1016/j.microc.2021.106012.

Kim, H. J., Kim, Y., Park, S. J., Kwon, C., & Noh, H. (2018). Development of Colorimetric Paper Sensor for Pesticide Detection Using Competitive-inhibiting Reaction. *Biochip Journal, 12*(4), 326–331. https://doi.org/10.1007/s13206-018-2404-z.

Kote, D. (2018). Increasing Food Safety Through the use of Information Technology. *European Journal of Engineering Research and Science, 3*(1), 35. https://doi.org/10.24018/ejers.2018.3.1.589.

Kröger, S., Setford, S. J., & Turner, A. P. F. (1998). Immunosensor for 2,4-dichlorophenoxyacetic acid in aqueous/organic solvent soil extracts. *Analytical Chemistry, 70*(23), 5047–5053. https://doi.org/10.1021/ac9805100.

Kumar, N., Krishnani, K. K., & Singh, N. P. (2019). Oxidative and Cellular Metabolic Stress of Fish: An Appealing Tool for Biomonitoring of Metal Contamination in the Kolkata Wetland, a Ramsar Site. *Archives of Environmental Contamination and*

Toxicology, *76*(3), 469–482. https://doi.org/10.1007/S00244-018-00587-5/METRICS.

Kumar, P., Kim, K. H., & Deep, A. (2015). Recent advancements in sensing techniques based on functional materials for organophosphate pesticides. In *Biosensors and Bioelectronics* (Vol. 70, pp. 469–481). Elsevier Ltd. https://doi.org/10.1016/j.bios.2015.03.066.

Kumar, T. H. V., & Sundramoorthy, A. K. (2019). Electrochemical biosensor for methyl parathion based on single-walled carbon nanotube/glutaraldehyde crosslinked acetylcholinesterase-wrapped bovine serum albumin nanocomposites. *Analytica Chimica Acta*, *1074*, 131–141. https://doi.org/10.1016/j.aca.2019.05.011.

Kumaravel, A., Murugananthan, M., Mangalam, R., & Jayakumar, S. (2020). A novel, biocompatible and electrocatalytic stearic acid/nanosilver modified glassy carbon electrode for the sensing of paraoxon pesticide in food samples and commercial formulations. *Food Chemistry*, *323*. https://doi.org/10.1016/j.foodchem.2020.126814.

Li, C., Wang, Z., & Zhan, G. (2011). Electrochemical investigation of methyl parathion at gold-sodium dodecylbenzene sulfonate nanoparticles modified glassy carbon electrode. *Colloids and Surfaces B: Biointerfaces*, *82*(1), 40–45. https://doi.org/10.1016/j.colsurfb.2010.08.011.

Li, C., Zhang, G., Wu, S., & Zhang, Q. (2018). Aptamer-based microcantilever-array biosensor for profenofos detection. *Analytica Chimica Acta*, *1020*, 116–122. https://doi.org/10.1016/j.aca.2018.02.072.

Li, H., Wang, Y., Zha, H., Dai, P., & Xie, C. (2019). Reagentless Electrochemiluminescence Sensor for Triazophos Based on Molecular Imprinting Electropolymerized Poly(Luminol-p-Aminothiophenol) Composite-Modified Gold Electrode. *Arabian Journal for Science and Engineering*, *44*(1), 145–152. https://doi.org/10.1007/s13369-018-3289-9.

Li, J., Lu, J., Qiao, X., & Xu, Z. (2017a). A study on biomimetic immunoassay-capillary electrophoresis method based on molecularly imprinted polymer for determination of trace trichlorfon residue in vegetables. *Food Chemistry*, *221*, 1285–1290. https://doi.org/10.1016/j.foodchem.2016.11.028.

Li, J., Lu, J., Qiao, X., & Xu, Z. (2017b). A study on biomimetic immunoassay-capillary electrophoresis method based on molecularly imprinted polymer for determination of trace trichlorfon residue in vegetables. *Food Chemistry*, *221*, 1285–1290. https://doi.org/10.1016/j.foodchem.2016.11.028.

Li, X., Gao, X., Gai, P., Liu, X., & Li, F. (2020). Degradable metal-organic framework/methylene blue composites-based homogeneous electrochemical strategy for pesticide assay. *Sensors and Actuators, B: Chemical*, *323*. https://doi.org/10.1016/j.snb.2020.128701.

Li, Y., Shi, L., Han, G., Xiao, Y., & Zhou, W. (2017). Electrochemical biosensing of carbaryl based on acetylcholinesterase immobilized onto electrochemically inducing porous graphene oxide network. *Sensors and Actuators, B: Chemical*, *238*, 945–953. https://doi.org/10.1016/j.snb.2016.07.152.

Lim, I. I. S., Chandrachud, U., Wang, L., Gal, S., & Zhong, C. J. (2008). Assembly-disassembly of DNAs and gold nanoparticles: A strategy of intervention based on oligonucleotides and restriction enzymes. *Analytical Chemistry*, *80*(15), 6038–6044. https://doi.org/10.1021/AC800813A.

Lisa, M., Chouhan, R. S., Vinayaka, A. C., Manonmani, H. K., & Thakur, M. S. (2009). Gold nanoparticles based dipstick immunoassay for the rapid detection of dichlorodiphenyltrichloroethane: An organochlorine pesticide. *Biosensors and Bioelectronics*, *25*(1), 224–227. https://doi.org/10.1016/j.bios.2009.05.006.

Liu, B., Zhou, P., Liu, X., Sun, X., Li, H., & Lin, M. (2013). Detection of Pesticides in Fruits by Surface-Enhanced Raman Spectroscopy Coupled with Gold Nanostructures. *Food and Bioprocess Technology*, *6*(3), 710–718. https://doi.org/10.1007/s11947-011-0774-5.

Liu, G., Guo, W., & Yin, Z. (2014). Covalent fabrication of methyl parathion hydrolase on gold nanoparticles modified carbon substrates for designing a methyl parathion biosensor. *Biosensors and Bioelectronics*, *53*, 440–446. https://doi.org/10.1016/j.bios.2013.10.025.

Liu, G., Li, L., Huang, X., Zheng, S., Xu, D., Xu, X., Zhang, Y., & Lin, H. (2018). Determination of triazole pesticides in aqueous solution based on magnetic graphene oxide functionalized MOF-199 as solid phase extraction sorbents. *Microporous and Mesoporous Materials*, *270*, 258–264. https://doi.org/10.1016/j.micromeso.2018.05.023.

Liu, Q., Fei, A., Huan, J., Mao, H., & Wang, K. (2015). Effective amperometric biosensor for carbaryl detection based on covalent immobilization acetylcholinesterase on multiwall carbon nanotubes/graphene oxide nanoribbons nanostructure. *Journal of Electroanalytical Chemistry*, *740*, 8–13. https://doi.org/10.1016/j.jelechem.2014.12.037.

Liu, Y., Pan, X., & Li, J. (2015). A 1961–2010 record of fertilizer use, pesticide application and cereal yields: a review. *Agronomy for Sustainable Development*, *35*(1), 83–93. https://doi.org/10.1007/s13593-014-0259-9.

Łozowicka, B., Rutkowska, E., & Jankowska, M. (2017). Influence of QuEChERS modifications on recovery and matrix effect during the multi-residue pesticide analysis in soil by GC/MS/MS and GC/ECD/NPD. *Environmental Science and Pollution Research*, *24*(8), 7124–7138. https://doi.org/10.1007/s11356-016-8334-1.

Madianos, L., Tsekenis, G., Skotadis, E., Patsiouras, L., & Tsoukalas, D. (2018). A highly sensitive impedimetric aptasensor for the selective detection of acetamiprid and atrazine based on microwires formed by platinum nanoparticles. *Biosensors and Bioelectronics*, *101*, 268–274. https://doi.org/10.1016/j.bios.2017.10.034.

Mallat, E., Barceló, D., Barzen, C., Gauglitz, G., & Abuknesha, R. (2001). Immunosensors for pesticide determination in natural waters. *TrAC - Trends in Analytical Chemistry*, *20*(3), 124–132. https://doi.org/10.1016/S0165-9936(00)00082-0.

Manav, Ö. G., Dinç-Zor, Ş., & Alpdoğan, G. (2019). Optimization of a modified QuEChERS method by means of experimental design for multiresidue determination of pesticides in milk and dairy products by GC–MS. *Microchemical Journal*, *144*, 124–129. https://doi.org/10.1016/j.microc.2018.08.056.

Mehta, J., Dhaka, S., Paul, A. K., Dayananda, S., & Deep, A. (2019). Organophosphate hydrolase conjugated UiO-66-NH2 MOF based highly sensitive optical detection of methyl parathion. *Environmental Research*, *174*, 46–53. https://doi.org/10.1016/j.envres.2019.04.018.

Mello, L. D., & Kubota, L. T. (2002). Review of the use of biosensors as analytical tools in the food and drink industries. In *Food Chemistry* (Vol. 77, Issue 2, pp. 237–256). https://doi.org/10.1016/S0308-8146(02)00104-8.

Moreno, J. L. F., Frenich, A. G., Bolaños, P. P., & Vidal, J. L. M. (2008). Multiresidue method for the analysis of more than 140 pesticide residues in fruits and vegetables by gas chromatography coupled to triple quadrupole mass spectrometry. *Journal of Mass Spectrometry*, *43*(9), 1235–1254. https://doi.org/10.1002/jms.1400.

Moro, L., Pezzotti, G., Turemis, M., Sanchís, J., Farré, M., Denaro, R., Giacobbe, M. G., Crisafi, F., & Giardi, M. T. (2018). Fast pesticide pre-screening in marine environment using a green microalgae-based optical bioassay. *Marine Pollution Bulletin*, *129*(1), 212–221. https://doi.org/10.1016/j.marpolbul.2018.02.036.

Mosiello, L., Laconi, C., Del Gallo, M., Ercole, C., & Lepidi, A. (2003). Development of a monoclonal antibody based potentiometric biosensor for terbuthylazine detection. *Sensors and Actuators, B: Chemical*, *95*(1–3), 315–320. https://doi.org/10.1016/S0925-4005(03)00431-3.

Navarro, P., Pérez, A. J., Gabaldón, J. A., Núñez-Delicado, E., Puchades, R., Maquieira, A., & Morais, S. (2013). Detection of chemical residues in tangerine juices by a duplex immunoassay. *Talanta*, *116*, 33–38. https://doi.org/10.1016/j.talanta.2013.04.062.

Navrátilová, I., & Skládal, P. (2004). The immunosensors for measurement of 2,4-dichlorophenoxyacetic acid based on electrochemical impedance spectroscopy. *Bioelectrochemistry*, *62*(1), 11–18. https://doi.org/10.1016/j.bioelechem.2003.10.004.

Neufeld, D. S. G., Åkerson, N., & Barahona, D. (2015). Pesticide Analysis in Vegetables Using QuEChERS Extraction and Colorimetric Detection. *Virginia Journal of Science*, *66*(4), 427–438. https://digitalcommons.odu.edu/vjs/vol66/iss4/4/.

Nie, Y., Teng, Y., Li, P., Liu, W., Shi, Q., & Zhang, Y. (2018). Label-free aptamer-based sensor for specific detection of malathion residues by surface-enhanced Raman scattering. *Spectrochimica Acta - Part A: Molecular and Biomolecular Spectroscopy*, *191*, 271–276. https://doi.org/10.1016/j.saa.2017.10.030.

Nowicka, A. M., Kowalczyk, A., Stojek, Z., & Hepel, M. (2010). Nanogravimetric and voltammetric DNA-hybridization biosensors for studies of DNA damage by common toxicants and pollutants. *Biophysical Chemistry*, *146*(1), 42–53. https://doi.org/10.1016/J.BPC.2009.10.003.

Paliwal, S., Wales, M., Good, T., Grimsley, J., Wild, J., & Simonian, A. (2007). Fluorescence-based sensing of p-nitrophenol and p-nitrophenyl substituent organophosphates. *Analytica Chimica Acta*, *596*(1), 9–15. https://doi.org/10.1016/j.aca.2007.05.034.

Pang, W., Zhao, H., Kim, E. S., Zhang, H., Yu, H., & Hu, X. (2012). Piezoelectric microelectromechanical resonant sensors for chemical and biological detection. In *Lab on a Chip* (Vol. 12, Issue 1, pp. 29–44). Royal Society of Chemistry. https://doi.org/10.1039/c1lc20492k.

Parab, A. E., Mohanapriya, K., & Jha, N. (2020). Non enzymatic electrochemical detection of paraoxon methyl using zinc oxide graphene nanocomposite in water and food samples. *Materials Today: Proceedings*, *42*, 710–717. https://doi.org/10.1016/j.matpr.2020.11.146.

Petersen, L. R., Beard, C. B., & Visser, S. N. (2019). Combatting the increasing threat of vector-borne disease in the United States with a national vector-borne disease prevention and control system. *American Journal of Tropical Medicine and Hygiene*, *100*(2), 242–245. https://doi.org/10.4269/ajtmh.18-0841.

Purdešová, A., & Dömötorová, M. (2017). MSPD as sample preparation method for determination of selected pesticide residues in apples. *Acta Chimica Slovaca*, *10*(1), 41–46. https://doi.org/10.1515/acs-2017-0007.

Qian, G., Wang, L., Wu, Y., Zhang, Q., Sun, Q., Liu, Y., & Liu, F. (2009). A monoclonal antibody-based sensitive enzyme-linked immunosorbent assay (ELISA) for the analysis of the organophosphorous pesticides chlorpyrifos-methyl in real samples. *Food Chemistry*, *117*(2), 364–370. https://doi.org/10.1016/j.foodchem.2009.03.097.

Qin, Y., Wu, Y., Chen, G., Jiao, L., Hu, L., Gu, W., & Zhu, C. (2020). Dissociable photoelectrode materials boost ultrasensitive photoelectrochemical detection of organophosphorus pesticides. *Analytica Chimica Acta*, *1130*, 100–106. https://doi.org/10.1016/j.aca.2020.07.036.

Quirino, J. P., & Terabe, S. (1998). Exceeding 5000-fold concentration of dilute analytes in micellar electrokinetic chromatography. In *Science* (Vol. 282, Issue 5388, pp. 465–468). https://doi.org/10.1126/science.282.5388.465.

Raman Suri, C., Kaur, J., Gandhi, S., & Shekhawat, G. S. (2008). Label-free ultra-sensitive detection of atrazine based on nanomechanics. *Nanotechnology*, *19*(23). https://doi.org/10.1088/0957-4484/19/23/235502.

Ramón-Azcón, J., Valera, E., Rodríguez, Á., Barranco, A., Alfaro, B., Sanchez-Baeza, F., & Marco, M. P. (2008). An impedimetric immunosensor based on interdigitated microelectrodes (IDμE) for the determination of atrazine residues in food samples. *Biosensors and Bioelectronics*, *23*(9), 1367–1373. https://doi.org/10.1016/j.bios.2007.12.010.

Ramos, J. J., González, M. J., & Ramos, L. (2009). Comparison of gas chromatography-based approaches after fast miniaturised sample preparation for the monitoring of selected pesticide classes in fruits. *Journal of Chromatography A*, *1216*(43), 7307–7313. https://doi.org/10.1016/j.chroma.2009.05.013.

Rapini, R., Cincinelli, A., & Marrazza, G. (2016). Acetamiprid multidetection by disposable electrochemical DNA aptasensor. *Talanta*, *161*, 15–21. https://doi.org/10.1016/j.talanta.2016.08.026.

Raymundo-Pereira, P. A., Gomes, N. O., Shimizu, F. M., Machado, S. A. S., & Oliveira, O. N. (2021). Selective and sensitive multiplexed detection of pesticides in food samples using wearable, flexible glove-embedded non-enzymatic sensors. *Chemical Engineering Journal*, *408*. https://doi.org/10.1016/j.cej.2020.127279.

Ren, X., & Chen, L. (2015). Quantum dots coated with molecularly imprinted polymer as fluorescence probe for detection of cyphenothrin. *Biosensors and Bioelectronics*, *64*, 182–188. https://doi.org/10.1016/j.bios.2014.08.086.

Rong, L., Wu, X., Xu, J., Dong, F., Liu, X., … X. P.-F. A. &, & 2018, undefined. (2017). Simultaneous determination of three pesticides and their metabolites in unprocessed foods using ultraperformance liquid chromatography-tandem mass spectrometry.

Taylor & FrancisL Rong, X Wu, J Xu, F Dong, X Liu, X Pan, P Du, D Wei, Y ZhengFood Additives & Contaminants: Part A, 2018•Taylor & Francis, *35*(2), 273–281. https://doi.org/10.1080/19440049.2017.1398419.

Salmain, M., Fischer-Durand, N., & Pradier, C. M. (2008). Infrared optical immunosensor: Application to the measurement of the herbicide atrazine. *Analytical Biochemistry*, *373*(1), 61–70. https://doi.org/10.1016/j.ab.2007.10.031.

Samal, S., Mohanty, R. P., Mohanty, P. S., Giri, M. K., Pati, S., & Das, B. (2023). Implications of biosensors and nanobiosensors for the eco-friendly detection of public health and agro-based insecticides: A comprehensive review. *Heliyon*, *9*(5), e15848. https://doi.org/10.1016/j.heliyon.2023.e15848.

Saraji, M., Jafari, M. T., & Mossaddegh, M. (2016). Carbon nanotubes@silicon dioxide nanohybrids coating for solid-phase microextraction of organophosphorus pesticides followed by gas chromatography–corona discharge ion mobility spectrometric detection. *Journal of Chromatography A*, *1429*, 30–39. https://doi.org/10.1016/J.CHROMA.2015.12.008.

Shahtaheri, S. J., Faridbod, F., & Khadem, M. (2017). Highly Selective Voltammetric Sensor Based on Molecularly Imprinted Polymer and Carbon Nanotubes to Determine the Dicloran Pesticide in Biological and Environmental Samples. *Procedia Technology*, *27*, 96–97. https://doi.org/10.1016/j.protcy.2017.04.041.

Shamsipur, M., Yazdanfar, N., & Ghambarian, M. (2016). Combination of solid-phase extraction with dispersive liquid-liquid microextraction followed by GC-MS for determination of pesticide residues from water, milk, honey and fruit juice. *Food Chemistry*, *204*, 289–297. https://doi.org/10.1016/j.foodchem.2016.02.090.

Sivaperumal, P., Anand, P., & Riddhi, L. (2015). Rapid determination of pesticide residues in fruits and vegetables, using ultra-high-performance liquid chromatography/time-of-flight mass spectrometry. *Food Chemistry*, *168*, 356–365. https://doi.org/10.1016/j.foodchem.2014.07.072.

Song, D., Wang, Y., Lu, X., Gao, Y., Li, Y., & Gao, F. (2018). Ag nanoparticles-decorated nitrogen-fluorine co-doped monolayer MoS2 nanosheet for highly sensitive electrochemical sensing of organophosphorus pesticides. *Sensors and Actuators, B: Chemical*, *267*, 5–13. https://doi.org/10.1016/j.snb.2018.04.016.

Sroysee, W., Chunta, S., Amatatongchai, M., & Lieberzeit, P. A. (2019). Molecularly imprinted polymers to detect profenofos and carbofuran selectively with QCM sensors. *Physics in Medicine*, *7*(June), 1–15. https://doi.org/10.1016/j.phmed.2019.100016.

Stobiecka, M., Cieśla, J. M., Janowska, B., Tudek, B., & Radecka, H. (2007). Piezoelectric Sensor for Determination of Genetically Modified Soybean Roundup Ready® in Samples not Amplified by PCR. *Mdpi.ComM Stobiecka, JM Cieśla, B Janowska, B Tudek, H RadeckaSensors, 2007•mdpi.Com*, *7*, 1462–1479. https://www.mdpi.com/1424-8220/7/8/1462.

Sulaiman, N. S., Rovina, K., & Joseph, V. M. (2019). Classification, extraction and current analytical approaches for detection of pesticides in various food products. *Journal Fur Verbraucherschutz Und Lebensmittelsicherheit*, *14*(3), 209–221. https://doi.org/10.1007/s00003-019-01242-4.

Sun, X., Gong, Z., Cao, Y., & Wang, X. (2013). Acetylcholinesterase Biosensor Based on Poly (diallyldimethylammonium chloride)-multi-walled Carbon Nanotubes-graphene Hybrid Film. *Nano-Micro Letters*, *5*(1), 47–56. https://doi.org/10.1007/BF03353731.

Sun, Y., Wei, J., Zou, J., Cheng, Z., Huang, Z., Gu, L., Zhong, Z., Li, S., Wang, Y., & Li, P. (2021). Electrochemical detection of methyl-paraoxon based on bifunctional cerium oxide nanozyme with catalytic activity and signal amplification effect. *Journal of Pharmaceutical Analysis*, *11*(5), 653–660. https://doi.org/10.1016/j.jpha.2020.09.002.

Tadeo, J. L., Sánchez-Brunete, C., Albero, B., & García-Valcárcel, A. I. (2010). Application of ultrasound-assisted extraction to the determination of contaminants in food and soil samples. In *Journal of Chromatography A* (Vol. 1217, Issue 16, pp. 2415–2440). https://doi.org/10.1016/j.chroma.2009.11.066.

Tan, M. J., Hong, Z. Y., Chang, M. H., Liu, C. C., Cheng, H. F., Loh, X. J., Chen, C. H., Liao, C. D., & Kong, K. V. (2017). Metal carbonyl-gold nanoparticle conjugates for highly sensitive SERS detection of organophosphorus pesticides. *Biosensors and Bioelectronics*, *96*, 167–172. https://doi.org/10.1016/j.bios.2017.05.005.

Tan, Y., Ahmad, I., & Wei, T. X. (2015). Detection of parathion methyl using a surface plasmon resonance sensor combined with molecularly imprinted films. *Chinese Chemical Letters*, *26*(6), 797–800. https://doi.org/10.1016/j.cclet.2015.04.001.

Tang, W., Yang, J., Wang, F., Wang, J., & Li, Z. (2019). Thiocholine-triggered reaction in personal glucose meters for portable quantitative detection of organophosphorus pesticide. *Analytica Chimica Acta*, *1060*, 97–102. https://doi.org/10.1016/j.aca.2019.01.051.

Tao, C. J., Hu, J. Y., Li, J. Z., Zheng, S. S., Liu, W., & Li, C. J. (2009). Multi-residue determination of pesticides in vegetables by gas chromatography/ion trap mass spectrometry. *Bulletin of Environmental Contamination and Toxicology*, *82*(1), 111–115. https://doi.org/10.1007/S00128-008-9528-0.

Timofeeva, I., Shishov, A., Kanashina, D., Dzema, D., & Bulatov, A. (2017). On-line in-syringe sugaring-out liquid-liquid extraction coupled with HPLC-MS/MS for the determination of pesticides in fruit and berry juices. *Talanta*, *167*, 761–767. https://doi.org/10.1016/j.talanta.2017.01.008.

Topuz, S., Özhan, G., & Alpertunga, B. (2005). Simultaneous determination of various pesticides in fruit juices by HPLC-DAD. *Food Control*, *16*(1), 87–92. https://doi.org/10.1016/J.FOODCONT.2003.11.012.

Tudi, M., Ruan, H. D., Wang, L., Lyu, J., Sadler, R., Connell, D., Chu, C., & Phung, D. T. (2021). Agriculture development, pesticide application and its impact on the environment. *Mdpi.ComM Tudi, H Daniel Ruan, L Wang, J Lyu, R Sadler, D Connell, C Chu, DT PhungInternational Journal of Environmental Research and Public Health, 2021•mdpi.Com*, *18*, 1112. https://doi.org/10.3390/ijerph18031112.

Vaghela, C., Kulkarni, M., Haram, S., Aiyer, R., & Karve, M. (2018). A novel inhibition based biosensor using urease nanoconjugate entrapped biocomposite membrane for potentiometric glyphosate detection. *International Journal of Biological Macromolecules*, *108*, 32–40. https://doi.org/10.1016/j.ijbiomac.2017.11.136.

Valera, E., Muñiz, D., & Rodríguez, Á. (2010). Fabrication of flexible interdigitated μ-electrodes (FIDμEs) for the development of a conductimetric immunosensor for atrazine detection based on antibodies labelled with gold nanoparticles.

Microelectronic Engineering, *87*(2), 167–173. https://doi.org/10.1016/j.mee.2009.07.001.

Vasylieva, N., Ahn, K. C., Barnych, B., Gee, S. J., & Hammock, B. D. (2015). Development of an Immunoassay for the Detection of the Phenylpyrazole Insecticide Fipronil. *Environmental Science and Technology*, *49*(16), 10038–10047. https://doi.org/10.1021/acs.est.5b01005.

Vohra, T., Grover, H., Saxena, S., Verma, D. K., & Rameshwari, R. (2020). A Review on Nanoparticles Based Biosensors for Pesticide Detection in Water. *Importance & Applications of Nanotechnology*, 1–9. https://meddocsonline.org/ebooks/ebook-nanotechnology/a-review-on-nanoparticles-based-biosensors-for-pesticide-detection-in-water.pdf.

Wang, F. R., Lee, G. J., Haridharan, N., & Wu, J. J. (2018). Electrochemical Sensor Using Molecular Imprinting Polymerization Modified Electrodes to Detect Methyl Parathion in Environmental Media. *Electrocatalysis*, *9*(1), 1–9. https://doi.org/10.1007/s12678-017-0411-9.

Wang, R. H., Zhu, C. L., Wang, L. L., Xu, L. Z., Wang, W. L., Yang, C., & Zhang, Y. (2019). Dual-modal aptasensor for the detection of isocarbophos in vegetables. *Talanta*, *205*(December), 1–5. https://doi.org/10.1016/j.talanta.2019.06.094.

Wang, W., Xu, Y., Cheng, N., Xie, Y., Huang, K., & Xu, W. (2020). Dual-recognition aptazyme-driven DNA nanomachine for two-in-one electrochemical detection of pesticides and heavy metal ions. *Sensors and Actuators, B: Chemical*, *321*. https://doi.org/10.1016/j.snb.2020.128598.

Wang, X., Mub, Z., Liu, R., Pu, Y., & Yin, L. (2013). Molecular imprinted photonic crystal hydrogels for the rapid and label-free detection of imidacloprid. *Food Chemistry*, *141*(4), 3947–3953. https://doi.org/10.1016/j.foodchem.2013.06.024.

Wang, X., Tang, Q., Wang, Q., Qiao, X., & Xu, Z. (2014). Study of a molecularly imprinted solid-phase extraction coupled with high-performance liquid chromatography for simultaneous determination of trace trichlorfon and monocrotophos residues in vegetables. *Journal of the Science of Food and Agriculture*, *94*(7), 1409–1415. https://doi.org/10.1002/jsfa.6429.

Wei, Y., Zhu, Y. ying, & Wang, M. li. (2016). A facile surface-enhanced Raman spectroscopy detection of pesticide residues with Au nanoparticles/dragonfly wing arrays. *Optik*, *127*(22), 10735–10739. https://doi.org/10.1016/j.ijleo.2016.08.118.

World Health Organization. (2020). The Who Recommended Classification of Pesticides By Hazard and Guidelines To Classification 2019. *World Health Organization*, 1–98. https://doi.org/ISBN 978 92 4 151796 3.

Wu, C. C. (2017). Multiresidue method for the determination of pesticides in Oolong tea using QuEChERS by gas chromatography-triple quadrupole tandem mass spectrometry. *Food Chemistry*, *229*, 580–587. https://doi.org/10.1016/j.foodchem.2017.02.081.

Wu, S., Li, D., Wang, J., Zhao, Y., Dong, S., & Wang, X. (2017). Gold nanoparticles dissolution based colorimetric method for highly sensitive detection of organophosphate pesticides. *Sensors and Actuators, B: Chemical*, *238*, 427–433. https://doi.org/10.1016/j.snb.2016.07.067.

Wu, W., Li, Y., Song, P., Xu, Q., Long, N., Li, P., Zhou, L., Fu, B., Wang, J., & Kong, W. (2023). Metal-organic framework (MOF)-based sensors for exogenous contaminants in food: Mechanisms, advances, and prospects. *Trends in Food Science and Technology*, *138*(May), 238–271. https://doi.org/10.1016/j.tifs.2023.06.016.

Xu, Y., Zhang, W., Shi, J., Li, Z., Huang, X., Zou, X., Tan, W., Zhang, X., Hu, X., Wang, X., & Liu, C. (2020). Impedimetric aptasensor based on highly porous gold for sensitive detection of acetamiprid in fruits and vegetables. *Food Chemistry*, *322*(February). https://doi.org/10.1016/j.foodchem.2020.126762.

Yan, X., Li, H., Hu, T., & Su, X. (2017). A novel fluorimetric sensing platform for highly sensitive detection of organophosphorus pesticides by using egg white-encapsulated gold nanoclusters. *Biosensors and Bioelectronics*, *91*, 232–237. https://doi.org/10.1016/j.bios.2016.11.058.

Yande, L., Yuxiang, Z., Haiyang, W., Bing, Y., D, L. Y., X, Z. Y., & Y, W. H. (2016). Detection of pesticides on navel orange skin by surface-enhanced Raman spectroscopy coupled with Ag nanostructures. *Ijabe.OrgL Yande, Z Yuxiang, W Haiyang, Y BingInternational Journal of Agricultural and Biological Engineering, 2016•ijabe.Org*, *9*(2), 179–185. https://doi.org/10.3965/j.ijabe.20160902.1960.

Yang, J., Wang, H., Jiang, Y., Sun, Y., Pan, K., Lei, H., Wu, Q., Shen, Y., Xiao, Z., & Xu, Z. (2008). Development of an enzyme-linked immuno-sorbent assay (ELISA) method for carbofuran residues. *Molecules*, *13*(4), 871–881. https://doi.org/10.3390/molecules13040871.

Ye, T., Yin, W., Zhu, N., Yuan, M., Cao, H., Yu, J., Gou, Z., Wang, X., Zhu, H., Reyihanguli, A., & Xu, F. (2018). Colorimetric detection of pyrethroid metabolite by using surface moleculary imprinted polymer. *Sensors and Actuators, B: Chemical*, *254*, 417–423. https://doi.org/10.1016/j.snb.2017.07.132.

Yin, H., Ai, S., Xu, J., Shi, W., & Zhu, L. (2009). Amperometric biosensor based on immobilized acetylcholinesterase on gold nanoparticles and silk fibroin modified platinum electrode for detection of methyl paraoxon, carbofuran and phoxim. *Journal of Electroanalytical Chemistry*, *637*(1–2), 21–27. https://doi.org/10.1016/j.jelechem.2009.09.025.

Yu, X., & Yang, H. (2017). Pyrethroid residue determination in organic and conventional vegetables using liquid-solid extraction coupled with magnetic solid phase extraction based on polystyrene-coated magnetic nanoparticles. *Food Chemistry*, *217*, 303–310. https://doi.org/10.1016/j.foodchem.2016.08.115.

Zare, A. R., Ensafi, A. A., & Rezaei, B. (2019). An impedimetric biosensor based on poly(l-lysine)-decorated multiwall carbon nanotubes for the determination of diazinon in water and fruits. *Journal of the Iranian Chemical Society*, *16*(12), 2777–2785. https://doi.org/10.1007/s13738-019-01741-z.

Zeng, K., & Grimes, C. A. (2004). Threshold-crossing counting technique for damping factor determination of resonator sensors. *Review of Scientific Instruments*, *75*(12), 5257–5261. https://doi.org/10.1063/1.1819631.

Zhai, C., Guo, Y., Sun, X., Zheng, Y., & Wang, X. (2014). An acetylcholinesterase biosensor based on graphene-gold nanocomposite and calcined layered double hydroxide. *Enzyme and Microbial Technology*, *58–59*, 8–13. https://doi.org/10.1016/j.enzmictec.2014.02.004.

Zhang, C., Zhao, F., She, Y., Hong, S., Cao, X., Zheng, L., Wang, S., Li, T., Wang, M., Jin, M., Jin, F., Shao, H., & Wang, J. (2019). A disposable molecularly imprinted sensor based on Graphe@AuNPs modified screen-printed electrode for highly selective and sensitive detection of cyhexatin in pear samples. *Sensors and Actuators, B: Chemical, 284*, 13–22. https://doi.org/10.1016/j.snb.2018.12.075.

Zhang, J., Wang, C., Niu, Y., Li, S., & Luo, R. (2017). Electrochemical sensor based on molecularly imprinted composite membrane of poly(o-aminothiophenol) with gold nanoparticles for sensitive determination of herbicide simazine in environmental samples. *Sensors and Actuators, B: Chemical, 249*, 747–755. https://doi.org/10.1016/j.snb.2016.02.068.

Zhang, L. (2013). Self-assembly Ag nanoparticle monolayer film as SERS Substrate for pesticide detection. *Applied Surface Science*, *270*(April), 292–294. https://doi.org/10.1016/j.apsusc.2013.01.014.

Zhang, P., Ding, J., Hou, J., Zhao, L., Chen, Y., & Ding, L. (2017). Dynamic microwave assisted extraction coupled with matrix solid phase dispersion for the determination of chlorfenapyr and abamectin in rice by LC-MS/MS. *Microchemical Journal*, *133*, 404–411. https://doi.org/10.1016/j.microc.2017.04.006.

Zhang, Q., Xu, Q., Guo, Y., Sun, X., & Wang, X. (2016). Acetylcholinesterase biosensor based on the mesoporous carbon/ferroferric oxide modified electrode for detecting organophosphorus pesticides. *RSC Advances*, *6*(29), 24698–24703. https://doi.org/10.1039/C5RA21799G.

Zhao, F., Wang, L., Li, M., Wang, M., Liu, G., & Ping, J. (2023). Nanozyme-based biosensor for organophosphorus pesticide monitoring: Functional design, biosensing strategy, and detection application. *TrAC - Trends in Analytical Chemistry*, *165*(August), 1–5. https://doi.org/10.1016/j.trac.2023.117152.

Zhao, G., Zhou, B., Wang, X., Shen, J., & Zhao, B. (2021). Detection of organophosphorus pesticides by nanogold/mercaptomethamidophos multi-residue electrochemical biosensor. *Food Chemistry*, *354*(February). https://doi.org/10.1016/j.foodchem.2021.129511.

Zhao, H., Ji, X., Wang, B., Wang, N., Li, X., Ni, R., & Ren, J. (2015). An ultra-sensitive acetylcholinesterase biosensor based on reduced graphene oxide-Au nanoparticles-β-cyclodextrin/Prussian blue-chitosan nanocomposites for organophosphorus pesticides detection. *Biosensors and Bioelectronics*, *65*, 23–30. https://doi.org/10.1016/j.bios.2014.10.007.

Chapter 4

Challenges and Future Directions in Pesticide Detection

Reena Chauhan[1,*]
Savita Rani[2]
Indu Chopra[3]
and Pooja[4]
[1]Department of Entomology, CCSHAU, Hisar, India
[2]Department of Horticulture, CCSHAU, Hisar, India
[3]Division of Soil Science and Agricultural Chemistry, IARI, New Delhi, India
[4]Department of Chemistry, CCSHAU, Hisar, India

Abstract

The need to regularly monitor the degree of pesticide contamination in various environmental and biological matrices is growing continuously. Five basic techniques are employed during sample preparation for extraction. The precise and trustworthy techniques for screening pesticides in various environmental matrices include spectrophotometry, polarography, and other chromatographic methods, as well as mass spectrometry detectors. Low detection limits, unstable enzyme activity, complex matrices, and multi-residue approaches encompassing a significant number of pesticides under examination are limitations that still need to be improved upon to meet demand. The transition scanning speed constrains the application of tandem mass spectrometry. The complexity of pesticide active ingredients poses a significant barrier to the detection of all biomarkers; therefore, although biosensors can detect pesticides in food, they still require refinement for field applications.

* Corresponding Author's Email: reenavansh82@gmail.com

In: Insecticides
Editors: Archana Anokhe and Govindaraj Kamalam Dinesh
ISBN: 979-8-90134-022-6

Additionally, it is acknowledged that not all these analytes will behave flawlessly. Therefore, we need to consider process changes, such as the addition of screening procedures and alternatives to our current triple-quadrupole MS systems, to handle this. Anionic herbicides, such as glyphosate and glufosinate, are being detected using high-resolution mass detection techniques as a potential future application; however, challenges remain in finding a robust and reliable method. Similar issues need to be addressed with challenging matrices, such as spices, tea, and coffee. Hence, finding transitions for each new molecule introduced to a method and optimizing those transitions for the instrument in use requires considerable work. This chapter covers all challenges and future applications of trace-level pesticide detection.

Keywords: pesticides, residues, analytical techniques, instruments, matrix

Introduction

The primary concern of all the countries is securing the food demand of the rapidly growing human population on the earth, and some great scholars predict that it may increase to around ten billion by 2050. According to facts, there is a 97 million per year increase in world population (Saravi and Shokrzadeh, 2011). FAO roughly estimated that nearly 500 million people are undernourished (Anon, 1990). As the global population increases, there is an urgent need to enhance agricultural production to meet the growing needs. This situation has imposed a significant burden on agribusiness, necessitating increased crop production inputs (Kumar et al., 2021).

Plant protectants (pesticides) are an undeniable part of today's life. They help farmers increase their income by enhancing agricultural production, improving product quality, reducing yield losses, lowering production costs (Kumari and Chauhan, 2015), extending the shelf life of products, and storing them in warehouses (Damalas and Eleftherohorinos, 2011; Chu et al., 2020). The success of the Green and Grey Revolution in plants, the Yellow Revolution in oilseed crops, and the White and Blue Revolution in animals was also greatly facilitated by pesticides. If pesticides had not been used in agriculture, the world's crop loss would have reached roughly 40 percent. For instance, India loses around 90,000 crores in agricultural produce due to infestations caused by insect pests (Bahri, 2018). According to a guesstimate, every penny spent on chemical pest management can save three crops (Jeyanthi and Kombairaju, 2005). In recent years, pesticide residues have been rejected due to contamination in agricultural produce exported from India to European Countries (Mukherjee et al., 2019). Due to the presence of these added contaminants (pesticide residues) in food commodities, it is a health concern, and authorities have paid attention to this matter. Low levels of residual contamination may also lead to diseases, *such as low immunity, slow learning capacity,* and a host of other ailments (Dudwal et al., 2022). The quality of foodstuffs that humans consume is an important and serious issue due to increased pesticide usage (FAO, 2022). The identification and quantification of these pesticide residues in various food matrices are carried out using analytical methods worldwide (Hajslova and Zrostlikova, 2003).

What Are Pesticides?

Pesticides are chemical-based products that humans knowingly add to the environment to save crops from infestation caused by harmful insect pests. Pesticide is a broad term that is further subdivided into several forms, including herbicides used to control weeds in the field, fungicides used to control fungal infections, rodenticides used to control rodent attacks, insecticides used to control insect infestations, plant growth regulators, and many others.

Importance of Pesticide Detection

There is no doubt that considering pesticides as a tool of modern agricultural farming is beneficial, as they help farmers increase their agricultural production and protect crops from damage. As we know, "each coin has two sides. "On one side, it raises the income of the farmers by enhancing the yield. On the other hand, these chemical-based pesticides can prove a curse if sprayed without following proper recommendations. They may leave residues on the edible portion of plants. This ultimately causes severe health issues (Dudwal et al., 2022). To reduce the level of risks to our environment and the health of living beings, residues must be within the permissible limits. The government of India is continuously working in this context by educating the farmers and passing laws for the safer and proper use of pesticides. Therefore, there is an urgent need to determine the residues of pesticides in various food commodities for the betterment of consumers, producers, and authorities responsible for food quality control (Aguilear et al., 2003).

Extraction and Cleanup Techniques

For extracting the residues of pesticides from the sample/matrix, the process of separation is known as extraction. The method employed for extraction must be appropriate, and the use of solvent should be as per the method or the nature of the pesticide, ensuring the quantitative removal of pesticide residues from the matrix. It does not change the basic chemical behaviour of pesticides. During extraction, it is necessary to use properly washed and sterilized glassware. Generally, three steps are involved in estimating pesticide residues in any matrix.

a) Extraction
b) Cleanup
c) Quantification

Till now, no single method is available for extracting the residues. Based on the nature and chemical properties of pesticides, the authors used different extraction procedures to evaluate their residual quantities (Syrgabek and Alimzhanova, 2022). Pesticides are a heterogeneous group of compounds with dissimilar biological and physicochemical properties. The first and most basic step is sample preparation, marked as "the most critical" one. Several pre-treatment and extraction techniques are available and in use for determining pesticide residues. Here are some extraction techniques that are briefly discussed:

Liquid-Liquid Extraction (LLE): Among the various extraction methods, LLE is the most traditional one, derived from the Luke method (1975). This method utilizes two dissimilar, immiscible organic solvents and mixes them to extract the residues (Figure 1). Various extraction solutions, such as acetone, dichloromethane, acetonitrile, ethyl acetate, cyclohexane, and hexane, are applied for pesticide determination (Rajput et al., 2021).

Soxhlet extraction: This technique was developed by Franz Ritter Von in 1879. In this procedure, solvent circulates through the extractor for a prolonged period, approximately 8-24 hours (Figure 2). This technique can extract dry materials, such as soil, cereal flour, and ground plant material. The ground sample is filled into a thimble made from Whatman filter paper and placed in a Soxhlet extractor. The solvent is then added to the extractor and placed on a heating mantle to extract the residues from the ground samples. Although extraction is most efficient, the formation of fine capillary tubes in the sample mass sometimes obstructs complete extraction.

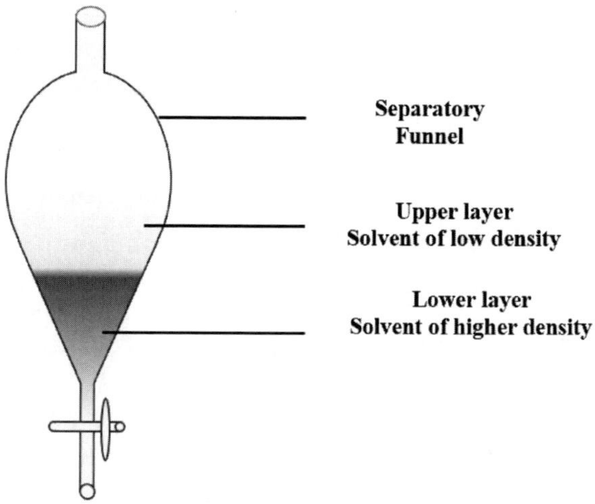

Figure 1. Process of Liquid-Liquid Extraction.

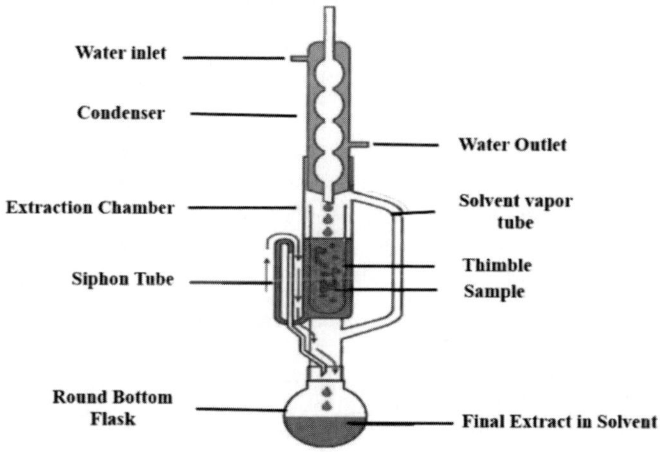

Figure 2. Process of Soxhlet extraction.

Extraction by blending: The material is chopped into small pieces and blended with a suitable solvent system in a warping blender, preferably explosion-proof, for 2 minutes. The method ensures close contact with the solvent and substrate; however, its significant disadvantage is the extraction of excessive co-extractives along with the toxicant.

Tumbling (Extraction by rotating or shaking with a solvent or mixture of solvents): This method primarily applies to raw fruit or vegetable crops with pesticide residues on the surface only. Processing is performed by chopping the plant material into pieces (approximately 0.5-1 cm), then placing it in an extractor, and the volume of solvent is generally kept equal to or twice the weight of the sample. Then, to establish proper contact between the sample and the solvent, the mixture is shaken for an hour.

Combination of tumbling and blending techniques: Here, the chopped sample is blended for 1-2 minutes with a minimum amount of extraction solvent, followed by tumbling for approximately an hour after the addition of the remaining solvents.

Limitations: The extraction procedures mentioned above are the most traditional methods that were used earlier. These methods have several limitations, including being time-consuming and laborious due to the multiple steps required for extracting residues from the matrix (Durovic R and Dorđevic T, 2011). Additionally, the organic solvents used are expensive and carcinogenic in nature. Inefficient extraction of polar pesticides, incomplete phase separation, and interference in extraction due to the formation of emulsions with aqueous samples (Pico et al., 2004). As a result, this technique is no longer used.

Several simple and rapid methods have been introduced to analyze the residues of pesticides. A few of them were discussed below:

Solid-liquid extraction (SLE): To extract the analytes from complex matrices, analytes are partitioned between a solid matrix and a water-miscible organic phase at standard temperature (Pinho et al., 2010), based on the distribution ratio of the analyte between solid and fluid (Zwir-Ferenc and Biziuk, 2004). This method is quite effective in extracting pesticides from plant-based materials (Avila et al., 2020). This method is highly advantageous due to the single steps involved in the extraction process; the requirement of solvents is very low, a cleanup step is not required, and it is a reliable process for extraction (Guedes et al., 2014).

Pressurized Liquid Extraction: Pressurized liquid extraction (PLE)is known by many names, such as accelerated solvent extraction (ASE), pressurized fluid extraction (PFE), pressurized hot solvent extraction (PHSE), high-pressure solvent extraction (HPSE), high-pressure high-temperature solvent extraction (HPHTSE), and subcritical solvent extraction (SSE) (Alvarez et al., 2020). The PLE technique was developed in 1995 (Hussein et al., 2007) for cost-cutting on solvent, with a reduced analysis time, and improved performance and high reproducibility of recovery compared to traditional extraction processes (Cho et al., 2008; Barriada et al., 2007). The technique is generally used for extracting the residues from the solid and semi-solid samples (Verma, 2010) at high temperatures and pressures. The Environmental Protection Agency (EPA) approved this method for determining the contaminants in environmental samples (Barp et al., 2023).

Microwave-assisted extraction (MAE): MAE is one of the fastest methods of extraction and is accepted by the US EPA method 3546 (Lahmanov and Varakina, 2019). This method was successfully used in determining pesticide residues in food commodities (Chung and Chen, 2011; Fuentes et al., 2009). MAE uses the energy of a microwave in the form of heat (Figure 3); the time required for extraction is less, as heat produced by the microwave is used to heat the whole sample instead of the container in which the sample was stored, resulting in the solution attaining its boiling point fast (Eskilsson and Björklund, 2000).

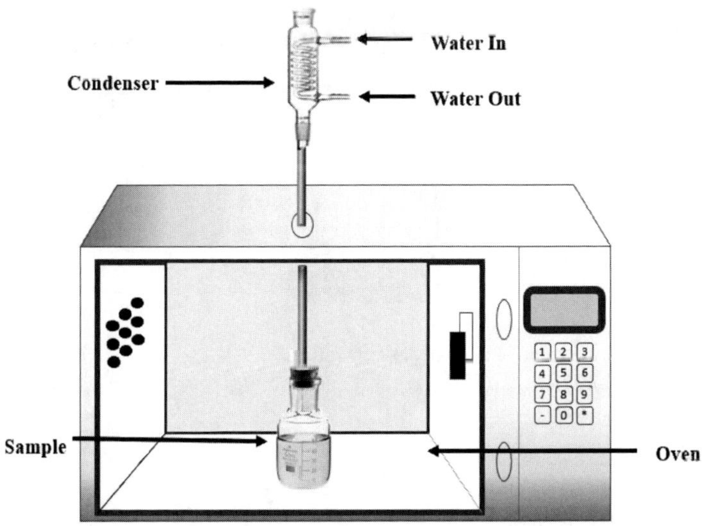

Figure 3. Microwave Assisted Extraction.

Solid-phase micro-extraction (SPME): In 1990, an innovative technique was developed by Arthur and Pawliszyn for sample preparation, which reduces the need for solvent or complicated apparatus for concentrating volatile or non-volatile compounds in liquid samples or headspace (Alpendurada, 2000). SPME has two steps; in the first step, the coated fibre is exposed to the sample or its headspace, and the target analytes partition from the sample matrix to the coating (Figure 4). In the second step, the fibre bearing the concentrated analytes is transferred to the analytical instrument, where the extracted analytes' desorption, separation, and quantification occur. In SPME, three modes are employed: direct extraction, headspace SPME, and membrane-protected SPME. In the direct extraction mode and headspace, an adsorbent polymer coats the fibre rod. The membrane-protected SPME is the most commonly used due to the positive effect of the selective-access membrane in protecting the coated polymer from interferences.

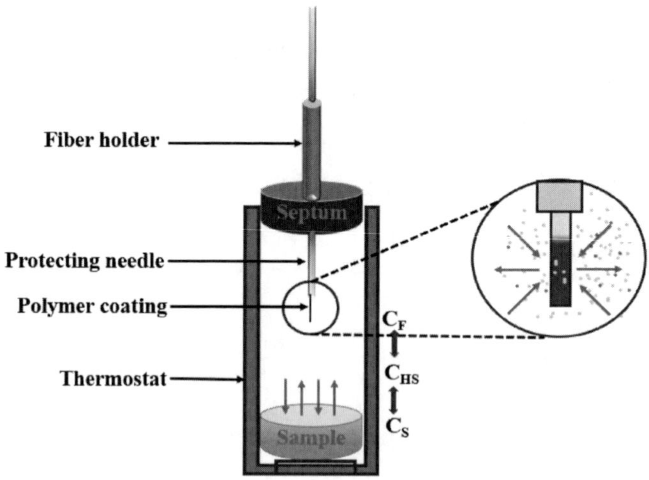

Figure 4. Process of Solid-Phase Microextraction.

Clean Up

One of the most important steps is to remove co-extractives and unwanted interfering substances, *such as waxes, pigments, and fats,* present in the extract, before analyzing the pesticide residues in the sample cleanup. Some of the commonly used cleanup methods are discussed herewith:

Steam Distillation

To remove natural oils, fats, or other steam-volatile co-extractives from non-volatile pesticides. Sometimes, volatile pesticides (Aldrin, Chlordane, Lindane) are removed from non-volatile extractives.

Partition

Polar pesticides can be removed from non-polar extracts using polar solvents. The conversion is also true. Pigments and waxes are preferentially soluble in n-hexane, whereas pesticides are extracted in acetonitrile due to their high polarity. This extract can be evaporated or diluted with water and re-extracted with hexane to recover the pesticides (Hoskin et al., 1958; Das, 2014).

Sweep Co-Distillation

Substances are extracted in a glass column packed with glass wool, followed by repeated injection of thc solvent. N_2 gas sweeps volatile compounds, including pesticides. Interfering substances remain in the column while the pesticide elutes out.

Coagulation

It is used for the cleanup of carbamate pesticides only. An aqueous NH_4Cl and phosphoric acid solution is added to the filtered solution to precipitate interfering substances, which are then removed by filtration.

Oxidation

To determine organophosphorus pesticides, they are oxidized to inorganic phosphate by a mixture of fuming nitric acid and perchloric acid. Inorganic phosphate is determined by the molybdenum blue method.

Saponification

Fats, oils, and triglyceride-containing samples can be easily cleaned by refluxing with alcoholic KOH, which converts triglycerides to soaps and glycerin.

Gel permeation chromatography (GPC) follows the principle of size exclusion, where molecules are separated based on their size (Dass, 2014). During the separation, molecules of large size are eluted first from the gel, followed by smaller molecules (Figure 5). In the 1970s, this method was used to extract and clean up pesticides (Martin et al., 2013). This method is the most effective among various cleanup methods for samples with complex matrices (Wu et al., 2011; Frenich et al., 2007; Pang et al., 2006). For filling the columns, polystyrene and divinylbenzene gel are used. Different combinations, such as ethyl acetate with cyclohexane, acetone with cyclohexane, and hexane with ethyl acetate, are used as mobile phases (Zwir-Ferenc and Biziuk, 2004).

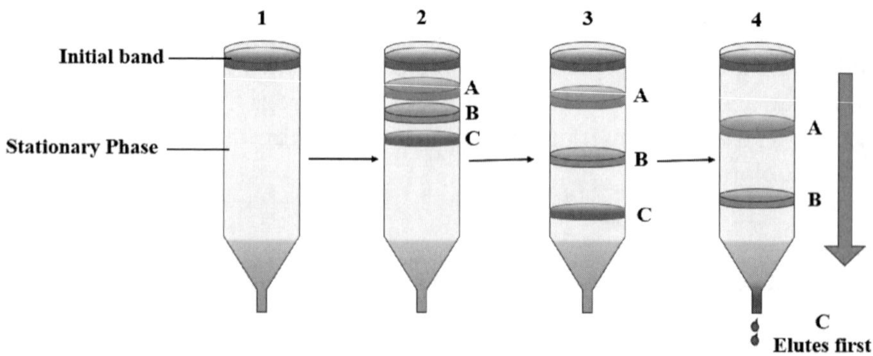

Figure 5. Process of gel permeation chromatography.

Solid-phase extraction (SPE): As a substitute, a new technique was developed in the 1970s to circumvent the problems of LLE, known as Solid-phase extraction (Thurman et al., 1998). SPE has gained worldwide acceptance in pesticide analysis (Figure 6). This process provides solventless extraction due to reduced occupational exposure risk, as well as more economical and faster extraction. (Aravinna and Alwis, 2003). It is based on the theory of liquid-solid phase chromatography, utilizing advanced sorbent technology that reorients with various functions depending on the structure of the extraction sorbents, including reversed-phase, normal-phase, cation exchange, anion exchange, and mixed-mode (Badawy et al., 2023).

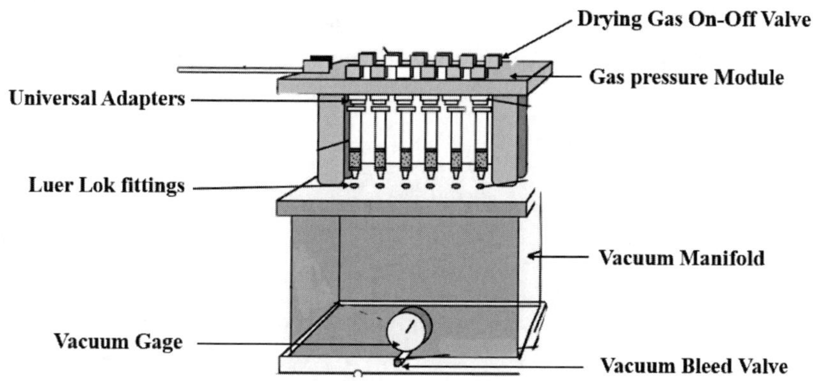

Figure 6. Solid-phase extraction.

Extraction Cum Clean Up

QuEChERS (Quick, easy, cheap, effective, rugged, and safe) method: The method was developed by Anastassiades et al., (2003) for analyzing the multi-pesticide residues (Figure 7). This technique can extract both polar and non-polar compounds simultaneously (Rajput et al., 2021). The technique promotes the aspect of green chemistry (Curyło et al., 2007; Lehotay et al., 2010) as it entails less solvent, requires a low amount of sample, produces toxic compounds almost entirely, does not require glassware, and is less energy- and time-consuming (Pszczolińska and Michel, 2016). The method is widely used for micro-scale extraction, and the results obtained are highly satisfactory (Farajzadeh and Dabbagh, 2020). The significant advantage of this method is its flexibility, i.e., one can modify the method according to the nature of the analyte and its composition (Fernandes et al., 2011). Acetonitrile is usually used as a solvent.

Matrix-solid-phase dispersion (MSPD): A new technique that was developed for the analysis of residues of pesticides (Martin et al., 2013) and is an elegant alternative to SPE that can be used for extraction and cleanup (Figure 8). This technique was developed by Barker et al., (1989), in which semi-solid, solid, and viscous samples can be directly handled, and interaction between the system and the components is much greater as compared to SPE (Lambropoulou and Albanis, 2007). It reduces the sample size and solvent volume and significantly works even if the sample amount is low (Kristenson et al., 2004). Common sorbents used as solid phases are C-8, C-18 (Grimalt and Dehouck, 2016), Florisil, or silica (Rajput et al., 2021). This sample is mixed with the solid phase until a homogeneous paste is formed, and then eluted by placing the homogeneous paste in a column, similar to SPE.

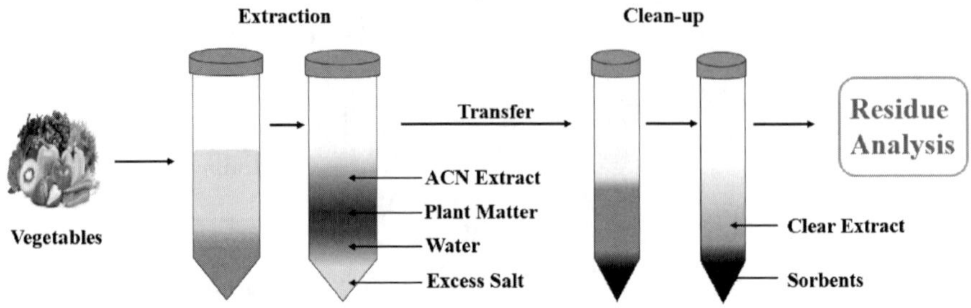

Figure 7. Diagrammatic representation of the QuEChERS method.

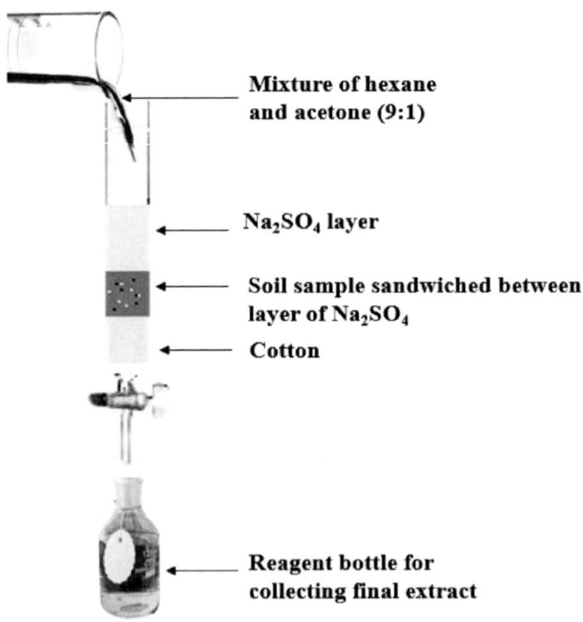

Figure 8. Process of matrix-solid phase extraction.

Table 1. Different multi-residue methods for the analysis of pesticide residues in different commodities

Sr. No.	Extraction Techniques	commodity	Pesticide	LOQ	LOD/Sensitivity of the instrument	Repeatability /Repeatability in %RSD	Instrument Used	Recovery (%)	References
1.	Solid-liquid extraction (SLE) methods with low-temperature partition	Husked rice Unhulled rice	Bifenthrin Deltamethrin Bifenthrin Deltamethrin	0.09 mg kg^{-1} 0.07 mg kg^{-1} 0.03 mg kg^{-1} 0.018 mg kg^{-1}	0.010 mg kg^{-1} 0.023 mg kg^{-1} 0.09 mg kg^{-1} 0.06 mg kg^{-1}	7	GC-ECD	< 95%	De Avila et al., 2020
2.	Solid phase extraction (SPE)	Water	alachlor, captan, chlorpyrifos, diazinon, profenophos, and oxyfluorfen.	0.09, 0.1, 0.004, 0.4, 0.04, 0.2 mg kg^{-1}	-	3	GC-ECD GC-NPD	80-120%	Aravinna and Alwis, 2003
3.	SPE	Water	18 pesticides	4.0-100.0 ng L^{-1}	0.4-40.0 ng L^{-1}	9	LC-ESI-MSMS	70-120%	Demoliner et al., 2010
4.	SPE with cartridge C-18	Ground water	Carbofuran, clomazone, tebuconazole, and 2,4-D	0.04 and 0.05 mg L^{-1}	0.01 and 0.02 mg L^{-1}	3	HPLC-DAD	60.3% to 107.7%	Cladas et al., 2019
5.	SPE with cartridge C-18	Groundwater	34 multiclass pesticides	0.017-0.099 µg/L	0.005-0.030 µg L^{-1}	3	GC-MS LC-MS	75-104%	Kapsi et al.,2020
6.	SPE	apples, cucumbers, red grapes, tomatoes	7 pesticides	0.1-0.5 mg kg^{-1}	-	0.3–4.1	LC-MS	90-104%	Oellig and Schwack, 2011

Table 1. (Continued)

Sr. No.	Extraction Techniques	commodity	Pesticide	LOQ	LOD/Sensitivity of the instrument	Repeatability /Repeatability in %RSD	Instrument Used	Recovery (%)	References
7.	SPE using new polymeric support (HLB and MCX) OASIS cartridges	human biological matrices	61 pesticides organophosphate, organochlorine, phthalimide, uracil carbamates, benzimidazoles	0.005 to 0-05 mg kg^{-1}	0.0025 to 0.02 ng ml^{-1}		GC-MS LC-MS	63 to 96%	Lacassie et al., 2001
8.	Soxhlet extractor SPE	Water sediments Water	18 organochlorine	0.006 µg/L 0.005 µg/L	0.001 µg/L 0.002 µg/L	3	GC-ECD	70 to 120 %	Saadati et al., 2013
9.	Solid phase microextraction (SPME)	Cowpea beans	Allyl isothiocyanate	0.33 µg kg^{-1}	0.11 µg kg^{-1}	$\leq$ 20%	GC-FID	108-114%	Vilela et al., 2020
10.	SPME	Mango Fruits	14 pesticide	3.33 to 33.33 microgram kg(-1)	1.0 to 3.3 microgram kg(-1)	3	GC-MS	71.6 to 117.5%	Menezes et al., 2010
11.	Liquid-liquid extraction (LLE) QuChERS	Chilli	Chlorantraniliprol e	0.01 mg kg-1	0.005 mg kg-1	5	GC-MSMS	72-91%	Ahlawat et al., 2019
12.	Pressurized liquid extraction (PLE)	Soil	14 Organochlorines	0.01 mg kg-1	-	3	GC-ECD	< 85%	Hussen et. al., 2007
13.	Microwave-assisted extraction (MAE) Soxhlet sonicator	bark, cement, grass, and vegetables	DDE, BHC, Dieldrin	-	-	3	GC-ECD	<100% for the grass and vegetables sample	Sanghi & Kannamkumarat (2004)

Sr. No.	Extraction Techniques	commodity	Pesticide	LOQ	LOD/Sensitivity of the instrument	Repeatability /Repeatability in %RSD	Instrument Used	Recovery (%)	References
14.	MAE	Grapes	8 pesticides	LOQ 2.2 to 5.8 μg kg^{-1}	LOD 0.7 to 1.7 μg kg^{-1}	2-8	GC-MS	82 to 107%	Lagunas-Allue et al., 2011
15.	Crude extracts, with no sample cleanup	grape, kiwi fruit, strawberry, spinach, lemon, peach, and nectarine	38 pesticides	0.005 mg/kg			LC-MSMS	60 to 140%	Taylor et al., 2002
16.	QuChERS	Pigeon Pea Soil	Chlorantraniliprole λ-cyhalothrin	0.004 mg/kg 0.003 for pigeon pea mg/kg for soil	0.001 mg/kg	5	UHPLC-MS/MS) GC-ECD	70 to 120%	Kansara et al., 2021
17.	QuChERS Cleanup Dispersive solid phase extraction (DSPE)	Three Types of Soil	10 OP Buprofezin thiadiazine	LOQ 8.64-53.7 μg/L for OP 141-195 μg/L for buprofezin 43.5-62.7°C for thiadiazine	2.52-27.9 μg/L for OP 42.3-58.5 μg/L buprofezin 43.5-62.7 μg/L thiadiazine	5	GC-NPD	93-113%	Asensio et al., 2010
18.	QuChERS	Cabbage	Ethion	0.05 mg kg^{-1}	0.02 mg kg^{-1}	3	GC-FPD	70-120%	Sawant et al., 2022
19.	QuChERS Cleanup DSPE	lettuce	Flubendiamide	LOQ 0.032 mg/g	0.009 mg/g	7	LC-MSMS	70-120%	Kuchipudi et al., 2019
20.	QuChERS	Grapes	fludioxonil	0.01 mg/kg	0.005 mg/kg	5	HPLCMSMS	81-94%	Yao et al., 2021
21.	QuChERS	74 Leafy vegetables	260 pesticides	1.88- 9.99 μg kg^{-1}	0.56 to 2.99 μg kg^{-1}	5	LC-MSMS	74.02–118.83%	Balkan and Yilmaz et al., 2022

Table 1. (Continued)

Sr. No.	Extraction Techniques	commodity	Pesticide	LOQ	LOD/Sensitivity of the instrument	Repeatability /Repeatability in %RSD	Instrument Used	Recovery (%)	References
22.	QuChERS without cleanup	cabbage and spinach	12 acidic pesticides	1–5 ng/g	-	<20%	LC-MSMS	95 to 110%	Islam et al., 2021
23.	QuChERS	200 Leafy vegetables	363 pesticides	0.01 mg kg^{-1}	0.05 mg kg^{-1}	5	LC-MSMS for 311 pesticides GC-MSMS for 52 pesticides	73 to116%	Deveci et al., 2023
24.	QuChERS Cleanup DSPE	Vine leaves	59 pesticides	LOQ mg kg^{-1} for 53 pesticides mg kg^{-1} for 5pesticides mg kg^{-1} for one pesticide	-	<20%	GC-MSMS	60-110%	Maestroni et al., 2018
25.	QuChERS micro-liquid-liquid extraction (MLLE)	Green Onion	spinetoram and spinosad	0.01 mg kg^{-1}		<12.5%	LC-MS/MS	78.3–93.4%	Malhat & Abdallah, 2019
26.	Modified QuChERS	wheat grains	25 multiclass pesticides	0.004 to 0.2 µg/g	0.002-0.06 µg/g	3	GC-ECD GC-MS	79.77-128.04%	Saini et al., 2016

Sr. No.	Extraction Techniques	commodity	Pesticide	LOQ	LOD/Sensitivity of the instrument	Repeatability /Repeatability in %RSD	Instrument Used	Recovery (%)	References
27.	QuEChERS	Fruits (grapes, cherries, peaches, and apricots) and vegetables (tomatoes, cucumbers, peppers, spinach, and zucchinis)	71 different pesticides and metabolites	0.40 to 7.21 µg kg^{-1}	0.12 to 2.16 µg kg^{-1}	Less than 20%	ultrahigh-performance liquid chromatography coupled to tandem mass spectrometry (UPLC/MS/MS)	70-120%	Bakırcı, and Hışıl 2012
28.	QuEChERS	grape and wine	160 pesticides	≤10 µg/L 12.5–25 µg/L (9 pesticides)	-	Less than 20%	GC × GC–time-of-flight mass spectrometric (TOFMS) method	70-120%	Dasgupta et al., 2010
29.	QuChERS micro-liquid-liquid extraction (MLLE)	Green Onion	spinetoram and spinosad	0.01 mg kg^{-1}		<12.5%	LC-MS/MS	78.3–93.4%	Malhat & Abdallah, 2019
30.	Microextraction No cleanup was necessary	grapes, must, and wine are used in vineyards	17 fungicides	0.01 to 0.05 mg/kg	0.007–0.005 mg kg^{-1}	7	GC-ECD GC-NPD	78 to 107%	Navarro et al., 2000
31.	Solid-liquid extraction (SLE)	Fruits and vegetables	12 insecticides	0.01 mg kg^{-1}	0.1-0.01 mg kg^{-1}	0.1-12.2%	HPLC-DAD Atmospheric pressure chemical ionization mass spectrometry (APCI-MS)	73 to 115%	Caboni et al., 2005

Table 1. (Continued)

Sr. No.	Extraction Techniques	commodity	Pesticide	LOQ	LOD/Sensitivity of the instrument	Repeatability /Repeatability in %RSD	Instrument Used	Recovery (%)	References
32.	Matrix solid-phase dispersion (MSPD)	8 different Fruits and vegetables	organophosphates	0.2–0.5 µg kg-1	0.06 to 0.15 µg kg^{-1}	1.8-11.8%	Rapid resolution liquid chromatography –tandem mass spectrometry (RRLC–MS–MS)	71.2-102.8%,	Guan et al., 2011
33.	MSPD	Eggplant, capsicum, apple gourd, cauliflower, sponge gourd	5 pesticides (Fungicides)	3.73-5.36 µg/mL	1.12-1.61 µg/mL	<20%	HPLC-UV Vis detector	88-116%	Mujahid et al., 2022

A brief overview of the different analytical methods used to extract residues of pesticides from various commodities, as summarized in published articles, is presented in Table 1. The table includes the limit of quantification (LOQ), limit of detection (LOD), repeatability, percentage recovery, and the instrument used for analysis.

Pesticide Residue Detection Techniques

The second strategy in pesticide analytical estimating uses several conventional analysis methods, such as the detection or analysis of target analytes (pesticides) in food. These methods include GC, HPLC, and more specialized ones, such as LC-MS/MS, GC-MS/MS, and ultra-high-performance liquid chromatography-tandem mass spectrometry (UHPLC-MS/MS). It is challenging to comprehend how to identify pesticides in actual samples due to the influence of various matrices. Due to their sensitivity, ability to separate substances, and ability to identify substances, gas and liquid chromatography have become the most popular methods for identifying and quantifying pesticides in recent years. In addition to this, several methods, including capillary electrophoresis (CE) and enzyme-linked immunosorbent assay (ELISA), as well as optical-based methods (Chen and Fung, 2018), have been used to identify pesticide residues.

Gas Chromatography (GC)

In multiple studies for volatile, easily vaporized insecticides, gas chromatography (GC) is the preferred isolation technology. It is frequently used when coupled with detection devices, such as FID, for the determination of pyrethroid pesticides in vegetable oils or organophosphorus pesticides in onions and grapes (Farajzadeh, 2016). Most studies reported indicate that GC is used in conjunction with different detectors to identify pesticides. Due to their sensitivity, mass selective detectors (MSD) (Łozowicka et al., 2016), nitrogen phosphorus detectors (NPD), flame photometric detectors (FPD) (Bakırcı et al., 2012), and electron capture detectors (ECD) are employed. Zhang, Gionfriddo, Acquaro, and Pawliszyn (2018) developed the direct immersion-solid-phase microextraction followed by gas chromatography-mass spectrometry (DISPME-GC-MS) method to identify PAHs, PCBs, and pesticide residues in seaweeds. By using gas chromatography-tandem mass spectrometry, Shabeer et al. (2018) developed a multi-residue approach to identify and quantify 243 pesticides in a cardamom matrix rapidly. In order to quantify folpet and phthalimide in food samples, Huertas-Perez, Ernest, Varela, and Badoud (2018) compared the effectiveness of GC paired with negative chemical ionization and GC-MSD and gas chromatography coupled with EI and GC-MS/MS. In the study, four different types of food matrices were used to validate both approaches, with GC-EI-MS demonstrating greater specificity and accuracy. As a result of polar pesticides' increased thermal instability and high volatility, GC approaches have lost some of their value in recent years.

Liquid Chromatography (LC)

Several liquid chromatography-based methods have been proposed for analyzing pesticide residues, with the majority incorporating UV, DAD, PDA, and MS detection devices. Octadecyl (C18), the most commonly used stationary phase, has been employed in gradient mode to reduce the time required for multi-residue analyses. A large number of techniques for identifying numerous pesticides in food, described in the scientific community, utilize RP-HPLC methodology, gradient mode, and an exponential increase in solvents based on the polarity of the pesticides. Only very seldom is an isocratic mode used. HPLC can be employed along with detectors like UV for pyrethroid residual evaluation in fruit crops or MSD/ MS-MS for triadimefon, imidacloprid, diazinon, and malathion detection in pineapple, apple, orange, and cherry (Bagheri et al., 2016; Timofeeva et al., 2017). To identify 60 pesticides in 286 samples of vegetables and fruits, time-of-flight mass spectrometry (TOF-MS) and ultra-high performance liquid chromatography (UHPLC) are now being employed. Another work used tandem mass spectrometry and UHPLC to identify 21 pesticides utilizing stationary-phase particle sizes that were significantly smaller than those used in typical LC (3-5 μm) (UHPLC-MS/MS). Castro et al., (2018) employed an integrated analysis technique that incorporated mass spectrometry, UPLC, and solid-phase extraction, enabling the detection of 50 pesticides with high precision. Yang, Zhao, and Hu (2018) created a multi-residue analysis method using ultra-performance liquid chromatography-triple quadrupole tandem mass spectrometry to assess and evaluate plant growth retardants. To achieve a smaller particle diameter, more powerful pressures must be applied. Due to the chromatographic system's significantly improved separation efficacy compared to normal HPLC, the UHPLC system is currently the preferred choice for multi-residue analysis.

Liquid Chromatography-Tandem Mass Spectrometry (LC-MS/MS)

One of the most cutting-edge analysis fields is the identification of pesticide residues using LC-MS. Almost every type of ionizing source and mass spectrometer has been used in this field. Most of these methods have been developed expressly for fruits and vegetables. To identify 18 chiral pesticides in water, soil, and river sediment samples, Zhao, Wang, Li, Guo, and Zhao (2018) devised a multi-residue analytical approach using magnetic solid-phase extraction (MAPE) and chiral LC-MS/MS. With the objective of testing four different herbicides (ethoxysulfuron, halosulfuron-methyl, mesosulfuron-methyl, and orthosulfamuron) in samples of cereal, Ni et al., (2018) evaluated a modified sample preparation approach followed by liquid chromatography-tandem mass spectrometry. Applications for LC-ESI MS/MS pesticide analysis in baby food are scarce. Quadrupole-time-of-flight systems (Q-TOF), quadrupole linear ion trap systems (Q-Trap), and triple quadrupole systems (QQQ) are the three most popular types of analyzers used in tandem mass spectrometry to detect pesticide residues in food (Halas, 2003).

Optical Screening Methods for Pesticide Residue in Food Matrices

The quest for straightforward, inexpensive, rapid, and on-site alternatives is necessary because chromatographic techniques are time-consuming, expensive, and require highly skilled personnel. Microplates have traditionally been used for biochemical tests that use antibodies or enzymes as identifying components; ELISA is a prime example of this type of bioassay. The technique uses specific antigen-antibody responses, making it sensitive and targeted. Since it moves quickly, it might be a helpful option when many samples need to be evaluated. Due to its simplicity (requires minimal training), low cost, and lack of expensive equipment, ELISA procedures can be seen as suitable for developing nations like India. ELISA has detected several Pesticides in a variety of food matrices. Duplex ELISA was used by Navarro et al., (2013) to identify the chlorpyrifos and fenthion residues in samples of tangerine juice. The suggested approach combined two different ELISA tests for the corresponding OPPs into a single ELISA test. The substantial advances in nanomaterials have further enhanced fluorescence detection.

Ambient Mass Spectrometry Methods

Condensed phase materials are employed in ambient mass spectrometry, which consists of a collection of processes that enable ions to be created from samples in an ambient environment before being captured and analyzed by mass spectrometry. This method is effective when testing for pesticide residues in food to ensure food safety. Desorption electrospray ionization (DESI) and real-time direct analysis were both pioneered by the rapidly growing field of ambient MS (Arrizabalaga et al., 2021; Sen et al., 2016). The most often used procedures include desorption electrospray ionization (DESI-MS), low-temperature plasma (LTP-MS), paper spray (PSMS), and direct analysis in real-time (DART-MS), all of which are commercially available. It has been demonstrated that analyzing fruit peels with DESI-MS as a DESI-MS substrate is a valuable screening method for examining materials containing pesticides (Beneito et al., 2020). Using DESI-MS, traces of atrazine were detected on Chinese cabbage leaves, dimethoate, tebuconazole, and trifloxystrobin on olive and vine leaves. The main restriction on quantitative surface analysis is the existence of matrix effects (Berchtold et al., 2013; Rocca et al., 2017). One of the most enticing features of ambient ionization sources is the use of portable mass spectrometers for live analysis.

Regulatory Consideration

Crop protection chemicals are a vital component of agriculture, as they play a major role in protecting crops and products from pests, thereby enhancing farm output. However, their indiscriminate usage has led to harmful residues in agricultural products, animal feed, agricultural soil, and drinking water. The Food Safety and Standards Act of 2006 regulates pesticide residue in food items. In contrast, the Insecticide Act of 1968 governs the registration of pesticides in India. The need to harmonize work on pesticide residues was felt due to variations in agricultural practices, analysis, and work plans, which led to inconsistent residue data outcomes. This led to the launch of the All India Coordinated Research Project (AICRP) on Pesticide Residues by the Indian

Council of Agricultural Research (ICAR) in 1984–85. Later, it was renamed as the All-India Network Project on Pesticide Residues (AINP-PR). Currently, there are 18 coordinated centres, with the Project Coordinating Cell located at the ICAR-Indian Agricultural Research Institute in New Delhi. Of these, 14 have been granted accreditation by the National Accreditation Board for Testing and Calibration of Laboratories (NABL) for conducting chemical testing in accordance with ISO/IEC 17025. The goal of the project is to ensure food safety and security by organizing, promoting, coordinating, and carrying out location-specific research on the analysis of pesticide residues in food commodities, soil, and water on a nationwide scale. Since 2005-06, the project has served as the central hub for implementing the 'Monitoring of Pesticide Residues at National Level' (MPRNL) Central Sector Scheme, supported by the Department of Agriculture and Farmers' Welfare (DA&FW).

Role of Regulatory Agencies in Pesticide Detection

Worldwide, many regulatory agencies (RAs) are working, which play an important role in the registration, approval, and ban of pesticides that are exported, imported, and marketed in large quantities globally. The pesticide industry relies heavily on regulatory agencies to ensure that the project plan accurately anticipates what the regulatory authority will require before approving the product. It acts as the interface between the regulatory authority and the project team, serving as the channel of communication with the regulatory authority as the project progresses. RA is responsible for staying up-to-date on relevant laws, rules, and other regulatory information. Regulatory bodies often expect businesses to use discretion when selecting how to apply such laws and recommendations. In India, the following two agencies are responsible for controlling pesticide use:

- Central Insecticides Board and Registration Committee (CIBRC)
- Food Safety and Standards Authority of India (FSSAI)

Central Insecticide Board and Registration Committee (CIBRC)

It is the supreme body where all the experts get together to decide whether a chemical can be registered in India or not. The novel chemical enters the Insecticide Schedule after receiving clearance from the CIB. The applicant may begin the registration procedure after the CIB approves the inclusion of the new molecule. To lawfully sell, trade, produce, or import the new molecule or its formulation in India once it has been placed in the insecticide schedule, the applicant must obtain registration of the new molecule under the different categories specified in Section 9(3). The registration committee (RC) facilitates the registration process through an online portal. The RC's numerous specialists investigate formulations to ensure they are effective and safe for both humans and animals, as claimed by the importer or manufacturer. Once all the conditions have been met, the RC will issue the applicant a certificate of registration (CR). In addition, the registration committee is responsible for conducting periodic reviews of all registered pesticides. It has the right to ban the use of such products that seem hazardous to human or animal health at any stage.

Food Safety and Standards Authority of India (FSSAI)

The FSSAI has established a Scientific Panel comprising subject matter experts to suggest Maximum Residue Levels (MRLs) in accordance with the Food Safety and Standards Act, 2006. Data on pesticide residues in India's registered crops, collected through supervised field trials at various locations and conducted in accordance with Good Agricultural Practices (GAP), are compiled by the State Agricultural Universities (SAUs)/Indian Council of Agricultural Research (ICAR). The Food Safety and Standards Authority of India (FSSAI), under the Ministry of Health and Family Welfare, gives due consideration to dietary exposure and risk assessment of pesticides. Maximum residual levels of pesticides on food crops, as established by the Food Safety and Standards Act of 2006, must be met before a pesticide can be approved for use.

Compliance with International Standards and Guidelines

There is considerable variation in international law regarding pesticides. When compared to underdeveloped countries, where necessary laws and policies are typically lacking, industrialized nations have stricter rules. Several developing nations struggle due to problems with policy implementation and enforcement, insufficient funds and personnel for conducting pesticide studies, and the requirement for specialized scientific knowledge (Handford et al., 2015)(Handford et al., 2015). The Codex Alimentarius Commission (CAC), an agency of the World Health Organization and the Food and Agriculture Organization, defines worldwide standards for food items and establishes limits on pesticide residues. Laying down a set of guidelines helps everyone involved in the food supply chain, at all levels, work together to reduce the likelihood of contamination and hazardous outcomes (Kubiak-Hardiman et al., 2023).

The Codex Maximum Residue Level (MRL) is the maximum allowable concentration of a pesticide in food or feed under current legislation, assuming the pesticide was administered correctly and in accordance with Good Agricultural Practices (GAP). After being recommended by the Codex Committee on Pesticide Residues (CCPR), CAC adopts them yearly as the worldwide voluntary standards. Three times a year, the CCPR meets. At the annual CAC meeting, based on the advice of an independent group of experts, the Joint Meeting on Pesticide Residues (JMPR) sets global standards for MRLs to allow free international trade of quality-compliant goods, protecting consumers' safety, and controlling daily intakes (Ambrus 2016). European pesticide law includes EU MRLs, which are often stricter than CODEX MRLs. In Europe, the European Food Safety Authority (EFSA) and the Member States review and approve new pesticides after a minimum three-year evaluation procedure. Global public health, as well as commerce and environmental interests, would benefit from harmonized regulations on pesticides.

Nonetheless, ensuring the food supply is safe for the public remains a persistent challenge. Risk management is a crucial tool that policymakers should utilize (Schierow 2012). To allow trade and sale of a product, a current MRL can be changed, especially if the product's residues may be higher than the national MRL (Kubiak-Hardiman et al., 2023).

Ultimately, global pesticide standards have been established to protect consumer health, preserve the natural environment, and facilitate the smooth flow of goods worldwide. Concerns about consumer safety and the efficacy of pesticide combinations vs. single pesticides are heightened by the fact that certain countries either do not have national MRLs or have MRLs that do not correlate with Codex MRLs. Disparities in pesticide regulation worldwide can sometimes lead to trade issues, particularly when developing countries use unauthorized pesticides or when maximum residue levels (MRLs) are not uniformly standardized.

Future Changes in Pesticide Regulation and Their Impact on Detection

As we know, CIBRC bans pesticides that seem to pose a threat to human health or the environment after conducting periodic reviews of registered pesticides. Still, it remains unclear whether the elimination of a particular pesticide will guarantee a reduction in public health risks or whether the nature and/or severity of the risk will shift due to the prohibition. Suppose a pesticide is banned to reduce one kind of danger. In that case, other pesticides may be introduced as replacements, potentially increasing the overall risk to public health. Therefore, before deciding to ban a pesticide, it is essential to conduct a risk assessment of both the current chemical in use and any potential alternatives, and predicting the balance between primary and secondary threats to human health is crucial. Suppose decision-makers anticipate that secondary and unintended risks will outweigh the main risk once a ban is implemented. In that case, they may put in place methods to mitigate them beforehand. Reducing application rates, limiting usage to specific crops, and raising the Maximum Containment Level (MCL) of alternative pesticides are all potential solutions. Increasing the use of alternative pesticides may alter the RfD and the percentage of exposure via other pathways, both of which factor into the MCL.

Conclusion and Future Prospects

Despite advancements in chromatographic systems' separation and detection, cleanup is still crucial for collecting accurate data. For the treatment of samples and the rapid identification of residual pesticides in fatty vegetable matrices, there remains a need for efficient, environmentally safe, and practical techniques. The development of modern extraction approaches will enable analysis at the low levels currently required by law for many pesticides, but more importantly, result in techniques that generate more trustworthy information to support food safety surveillance initiatives. The use of disposable chemicals or sorbents-only procedures, which involve low solvent consumption, such as QuEChERS, has superseded earlier techniques for pesticide multi-residue analysis. However, creating a special protocol that covers such a broad range of substances is challenging. Finally, large-scale multi-residue approaches targeting an extensive range of pesticides are highly desirable; however, the development of such procedures is complicated by the diverse types, classes, and physicochemical characteristics of pesticides.

Likewise, pesticides are typically detected using established analytical techniques, such as gas and liquid chromatography, which offer high selectivity and responsiveness at low detection limits. These procedures have a few drawbacks, though, including the fact that they are time-consuming, expensive, and labour-intensive, and require highly skilled personnel and expensive equipment. As a result, a sophisticated method for identifying pesticides, such as sensor-based technology, is being described. This method has several advantages over traditional chromatographic methods, including low cost, simple operation, high sensitivity and selectivity, on-site detection, and lower detection limits. The primary detection methods in biosensors include optical, electrochemical, piezoelectric, and molecularly imprinted polymers. The detection of pesticide residues in various sample matrices is being done using a few sensor platforms. Furthermore, the natural enzyme-like catalytic activity of nanomaterials, also known as nanozyme activity, has garnered considerable attention. A rapid and straightforward method for detecting pesticides in food products has been developed using nanotechnology.

Apart from the benefits offered by these detection methods, the main disadvantage was discovered to be the small number of pesticides detected, which renders the method vulnerable and necessitates the development of a good, affordable, and environmentally friendly method that could detect a large number of residual pesticides in a single run at lower limits. Technologies for sample preparation that are dependable, available, economical, and environmentally benign should be used in the future. Precision, selection, and speed are essential to enhance our analytical capabilities for these reasons and to stay one step ahead of restricted levels. Additionally, a quick analytical turnaround time is crucial for providing the necessary data for effective and timely residual pesticide evaluation in food items.

References

Ahammed Shabeer, T. P., Girame, R., Utture, S., Oulkar, D., Banerjee, K., Ajay, D., Arimboor, R., & Menon, K. R. K. (2018). Optimization of multi-residue method for targeted screening and quantitation of 243 pesticide residues in cardamom (Elettaria cardamomum) by gas chromatography tandem mass spectrometry (GC-MS/MS) analysis. *Chemosphere*, *193*, 447–453.

Ahlawat, S., Gulia, S., Malik, K., Rani, S., & Chauhan, R. (2019). Persistence and decontamination studies of chlorantraniliprole in Capsicum annum using GC–MS/MS. *Journal of food science and technology*, *56*, 2925-2931.

Ambrus, A. (2016). International harmonization of food safety assessment of pesticide residues. *Journal of agricultural and food chemistry*, *64*(1), 21-29.

Aravinna, A. G. P., & de Alwis, H. (2003). Determination of Pesticide Residues in Water by Solid Phase Extraction and GC/ECD, NPD. In *Proceedings of International Workshop on Technology of Application of Pesticides, June* (pp. 23-26).

Anon (1990a). Evidence of the bureau of rural resources to the senate select committee inquiry into the use of agricultural and veterinary chemicals in Australia.

Aguilera, A., Brotons, M., Rodríguez, M., & Valverde, A. (2003). Supercritical fluid extraction of pesticides from a table-ready food composite of plant origin (gazpacho). *Journal of agricultural and food chemistry*, *51*(19), 5616-5621.

Alvarez-Rivera, G., Bueno, M., Ballesteros-Vivas, D., Mendiola, J. A., & Ibanez, E. (2020). Pressurized liquid extraction. In *Liquid-phase extraction* (pp. 375-398). Elsevier.

Asensio-Ramos, M., Hernández-Borges, J., Ravelo-Pérez, L. M., & Rodriguez-Delgado, M. A. (2010). Evaluation of a modified QuEChERS method for the extraction of pesticides from agricultural, ornamental and forestal soils. *Analytical and Bioanalytical Chemistry*, *396*, 2307-2319.

Bakırcı, G. T., & Hışıl, Y. (2012). Fast and simple extraction of pesticide residues in selected fruits and vegetables using tetrafluoroethane and toluene followed by ultrahigh-performance liquid chromatography/tandem mass spectrometry. *Food chemistry*, *135*(3), 1901-1913.

Balkan, T., & Yılmaz, Ö. (2022). Method validation, residue and risk assessment of 260 pesticides in some leafy vegetables using liquid chromatography coupled to tandem mass spectrometry. *Food Chemistry*, *384*, 132516.

Bahri, C. (2018, October 27). New Migrant To India Could Threaten National (And Asian) Food Security. iS. India Spend. Retrieved from https://www.indiaspend.com/.

Barriada-Pereira, M., González-Castro, M. J., Muniategui-Lorenzo, S., López-Mahía, P., Prada-Rodríguez, D., & Fernández-Fernández, E. (2007). Comparison of pressurized liquid extraction and microwave assisted extraction for the determination of organochlorine pesticides in vegetables. *Talanta*, *71*(3), 1345-1351.

Barp, L., Višnjevec, A. M., & Moret, S. (2023). Pressurized Liquid Extraction: A Powerful Tool to Implement Extraction and Purification of Food Contaminants. *Foods*, *12*(10), 2017.

Bagheri, H., Yamini, Y., Safari, M., Asiabi, H., Karimi, M., & Heydari, A. (2016). Simultaneous determination of pyrethroids residues in fruit and vegetable samples via supercritical fluid extraction coupled with magnetic solid phase extraction followed by HPLC-UV. *The Journal of Supercritical Fluids*, *107*, 571–580.

Beneito-Cambra, M., Gilbert-López, B., Moreno-González, D., Bouza, M., Franzke, J., García-Reyes, J. F., & Molina-Díaz, A. (2020). Ambient (desorption/ionization) mass spectrometry methods for pesticide testing in food: a review. *Analytical Methods: Advancing Methods and Applications*, *12*(40), 4831–4852.

Berchtold, C., Müller, V., Meier, L., Schmid, S., & Zenobi, R. (2013). Direct detection of chlorpropham on potato skin using desorption electrospray ionization: Detection of chlorpropham by DESI. *Journal of Mass Spectrometry*, *48*(5), 587–593.

Caboni, P., Sarais, G., Angioni, A., Garau, V. L., & Cabras, P. (2005). Fast and versatile multiresidue method for the analysis of botanical insecticides on fruits and vegetables by HPLC/DAD/MS. *Journal of agricultural and food chemistry*, *53*(22), 8644-8649.

Caldas, S. S., Demoliner, A., & Primel, E. G. (2009). Validation of a method using solid phase extraction and liquid chromatography for the determination of pesticide residues in groundwaters. *Journal of the Brazilian Chemical Society*, *20*, 125-132.

Castro, G., Pérez-Mayán, L., Rodríguez-Cabo, T., Rodríguez, I., Ramil, M., & Cela, R. (2018). Multianalyte, high-throughput liquid chromatography tandem mass spectrometry method for the sensitive determination of fungicides and insecticides in wine. *Analytical and Bioanalytical Chemistry*, *410*(3), 1139–1150.

Chen, Q., & Fung, Y. (2010). Capillary electrophoresis with immobilized quantum dot fluorescence detection for rapid determination oforganophosphorus pesticides in vegetables. *Electrophoresis*, *31*, 3107–3114.

Chu, Y., Tong, Z., Dong, X., Sun, M., Gao, T., Duan, J., & Wang, M. (2020). Simultaneous determination of 98 pesticide residues in strawberries using UPLC-MS/MS and GC-MS/MS. *Microchemical Journal*, *156*, 104975.

Chung, S. W., & Chen, B. L. (2011). Determination of organochlorine pesticide residues in fatty foods: A critical review on the analytical methods and their testing capabilities. *Journal of Chromatography A*, *1218*(33), 5555-5567.

Damalas, C. A., & Eleftherohorinos, I. G. (2011). Pesticide exposure, safety issues, and risk assessment indicators. *International journal of environmental research and public health*, *8*(5), 1402-1419.

Dudwal, R., Jakhar, B. L., Pathan, A. R. K., Jan, I., Kakralya, B. L., Dhaka, S. R., &Coudhary, S. K. (2023). Dissipation kinetics, risk assessment, and waiting period of spiromesifen on chili fruits using gas chromatography–electron capture detector. *Biomedical Chromatography*, *37*(4), e5577.

Das, S. K. (2014). Recent developments in clean up techniques of pesticide residue analysis for toxicology study: a critical review. *Universal journal of agricultural research*, *2*(6), 198-202.

de Pinho, G. P., Neves, A. A., de Queiroz, M. E. L. R., & Silvério, F. O. (2010). Pesticide determination in tomatoes by solid–liquid extraction with purification at low temperature and gas chromatography. *Food chemistry*, *121*(1), 251-256.

Das gupta, S., Banerjee, K., Patil, S. H., Ghaste, M., Dhumal, K. N., & Adsule, P. G. (2010). Optimization of two-dimensional gas chromatography time-of-flight mass spectrometry for separation and estimation of the residues of 160 pesticides and 25 persistent organic pollutants in grape and wine. *Journal of Chromatography A*, *1217*(24), 3881-3889.

Deveci, B., Golge, O., & Kabak, B. (2023). Quantification of 363 Pesticides in Leafy Vegetables (Dill, Rocket and Parsley) in the Turkey Market by Using QuEChERS with LC-MS/MS and GC-MS/MS. *Foods*, *12*(5), 1034.

Demoliner, A., Caldas, S. S., Costa, F. P., Gonçalves, F. F., Clementin, R. M., Milani, M. R., & Primel, E. G. (2010). Development and validation of a method using SPE and LC-ESI-MS-MS for the determination of multiple classes of pesticides and metabolites in water samples. *Journal of the Brazilian Chemical Society*, *21*, 1424-1433.

de Ávila, M. B., Faroni, L. R., Heleno, F. F., de Queiroz, M. E. L., & Almeida, R. I. (2020). A simple and efficient optimized solid-liquid extraction with low temperature partition procedure for dissipation and translocation study of pesticide residues in rice grains. *Journal of the Brazilian Chemical Society*, *31*, 1991-1999.

Eskilsson, C. S., & Björklund, E. (2000). Analytical-scale microwave-assisted extraction. *Journal of chromatography A*, *902*(1), 227-250.

FAOSTAT. (2022). Food and Agriculture Data for All FAO Regional Groupings. Available online: https://www.fao.org/statistics/en (accessed on 17 April 2022).

Farajzadeh, M. A., Afshar Mogaddam, M. R., Rezaee Aghdam, S., Nouri, N., & Bamorrowat, M. (2016). Application of elevated temperature-dispersive liquid-liquid microextraction for determination of organophosphorus pesticides residues in aqueous samples followed by gas chromatography-flame ionizationdetection. *Food Chemistry*, *212*, 198–204.

Fuentes, E., Báez, M. E., & Díaz, J. (2009). Microwave-assisted extraction at atmospheric pressure coupled to different cleanup methods for the determination of organophosphorus pesticides in olive and avocado oil. *Journal of Chromatography A*, *1216*(51), 8859-8866.

Frenich, A. G., Bolaños, P. P., & Vidal, J. L. M. (2007). Multiresidue analysis of pesticides in animal liver by gas chromatography using triple quadrupole tandem mass spectrometry. *Journal of Chromatography A*, *1153*(1-2), 194-202.

Guan, S. X., Yu, Z. G., Yu, H. N., Song, C. H., Song, Z. Q., & Qin, Z. (2011). Multi-walled carbon nanotubes as matrix solid-phase dispersion extraction adsorbent for simultaneous analysis of residues of nine organophosphorus pesticides in fruit and vegetables by rapid resolution LC–MS–MS. *Chromatographia*, *73*, 33-41.

Guedes, T. D. J., Heleno, F. F., Amaral, M. D. O., Pinto, N. A., de Queiroz, M. E. L., Silva, D. F. D., & Silva, A. A. D. (2014). A simple and efficient method employing solid-liquid extraction with low-temperature partitioning for the determination/monitoring of pesticide residues in strawberries by GC/ECD. *Journal of the Brazilian Chemical Society*, *25*, 1520-1527.

Hajšlová, J., & Zrostlıkova, J. (2003). Matrix effects in (ultra) trace analysis of pesticide residues in food and biotic matrices. *Journal of Chromatography A*, *1000*(1-2), 181-197.

Handford, C. E., Elliott, C. T., & Campbell, K. (2015). A review of the global pesticide legislation and the scale of challenge in reaching the global harmonization of food safety standards. *Integrated environmental assessment and management, 11*(4), 525-536.

Hoskins, W. M., Erwin, W. R., Miskus, R., Thornburg, W. W., & Werum, L. N. (1958). Pesticide residue analysis, A polyethylene-alumina column for purification of tissue extracts before analysis. *Journal of Agricultural and Food Chemistry, 6*(12), 914-916.

Hussen, A., Westbom, R., Megersa, N., Mathiasson, L., & Björklund, E. (2007). Selective pressurized liquid extraction for multi-residue analysis of organochlorine pesticides in soil. *Journal of chromatography A, 1152*(1-2), 247-253.

Hałas, S., Spektrometria MasArrizabalaga-Larrañaga, A., Ayala-Cabrera, J., Seró, R., Santos, J., & Moyano, E. (2003). Ambient ionization mass spectrometry in food analysis. In *Food Toxicology and Forensics, 54*, 271–312.

Huertas-Perez, J. F., Ernest, M., Varela, J., & Badoud, F. (2018). Quantification of folpet and phthalimide in food by gas chromatography and mass spectrometry: Overcoming potential analytical artefacts. *Food Chemistry, 260*, 213–220.

Hussen, A., Westbom, R., Megersa, N., Mathiasson, L., & Björklund, E. (2007). Selective pressurized liquid extraction for multi-residue analysis of organochlorine pesticides in soil. *Journal of chromatography A, 1152*(1-2), 247-253.

Islam, A. K. M. M., Noh, H. H., Ro, J. H., Kim, D., Oh, M. S., Son, K., & Kwon, H. (2021). Optimization and validation of a method for the determination of acidic pesticides in cabbage and spinach by modifying QuEChERS procedure and liquid chromatography–tandem mass spectrometry. *Journal of Chromatography B, 1173*, 122667.

Jeyanthi, H., & Kombairaju, S. (2005). Pesticide use in vegetable crops: frequency, intensity and determinant factors. *Agricultural Economics Research Review, 18*(2), 209-221.

Kapsi, M., Tsoutsi, C., & Albanis, T. (2020). Simple analytical methodology based on solid phase extraction for monitoring pesticide residues in natural waters. *MethodsX, 7*, 101011.

Kuchipudi, A., Angappan, S., & Kaithamalai, B. (2019). Method Validation, Dissipation and Decontamination of Flubendiamide Residues in Lettuce using LC/MS/MS. *Pesticide Research Journal, 31*(1), 34-41.

Kansara, R. V., Solanki, V. H., Singh, S., & Chauhan, D. (2021). Persistence, dissipation behavior and dietary risk assessment of a combi-product of chlorantraniliprole and λ cyhalothrin in/on pigeonpea (Cajanus cajan L.). *Heliyon, 7*(4).

Kubiak-Hardiman, P., Haughey, S. A., Meneely, J., Miller, S., Banerjee, K., & Elliott, C. T. (2023). Identifying gaps and challenges in global pesticide legislation that impact the protection of consumer health: Rice as a case study. *Exposure and Health, 15*(3), 597-618.

Kumar, J., Ramlal, A., Mallick, D., & Mishra, V. (2021). An overview of some biopesticides and their importance in plant protection for commercial acceptance. *Plants, 10*(6), 1185.

Kumari, B., & Chauhan, R. (2015). Persistence and effect of processing on reduction of chlorpyriphos in chilli. *Journal of food processing and Technology, 6*(10).

Lahmanov, D. E., & Varakina, Y. I. (2019, April). A short review of sample preparation methods for the pesticide residue analysis in fatty samples. In *IOP Conference Series: Earth and Environmental Science* (Vol. 263, No. 1, p. 012061). IOP Publishing.

Lambropoulou, D. A., & Albanis, T. A. (2007). Methods of sample preparation for determination of pesticide residues in food matrices by chromatography–mass spectrometry-based techniques: a review. *Analytical and Bioanalytical Chemistry, 389*, 1663-1683.

Łozowicka, B.; Rutkowska, E.; Jankowska, M.(2017). Influence of QuEChERS modifications on recovery and matrix effect during the multi-residue pesticide analysis in soil by GC/MS/MS and GC/ECD/NPD. *Environ. Sci. Pollut. Res.*, *24*, 7124–7138.

Lagunas-Allue, L., Sanz-Asensio, J., & Martinez-Soria, M. T. (2011). Validation of a microwave-assisted extraction gas chromatography detection method for the determination of fungicides in grapes. *Analytical Methods*, *3*(12), 2881-2892.

Lacassie, E., Marquet, P., Gaulier, J. M., Dreyfuss, M. F., & Lachâtre, G. (2001). Sensitive and specific multiresidue methods for the determination of pesticides of various classes in clinical and forensic toxicology. *Forensic science international*, *121*(1-2), 116-125.

Martins, J. G., Chávez, A. A., Waliszewski, S. M., Cruz, A. C., & Fabila, M. M. G. (2013). Extraction and cleanup methods for organochlorine pesticides determination in milk. *Chemosphere*, *92*(3), 233-246.

Malhat, F., & Abdallah, O. (2019). Residue distribution and risk assessment of two macrocyclic lactone insecticides in green onion using micro-liquid-liquid extraction (MLLE) technique coupled with liquid chromatography tandem mass spectrometry. *Environmental monitoring and assessment*, *191*, 1-10.

Maestroni, B., Abu Alnaser, A., Ghanem, I., Islam, M., Cesio, V., Heinzen, H., & Cannavan, A. (2018). Validation of an analytical method for the determination of pesticide residues in vine leaves by GC-MS/MS. *Journal of agricultural and food chemistry*, *66*(25), 6421-6430.

Menezes Filho, A., dos Santos, F. N., & de Paula Pereira, P. A. (2010). Development, validation and application of a methodology based on solid-phase micro extraction followed by gas chromatography coupled to mass spectrometry (SPME/GC–MS) for the determination of pesticide residues in mangoes. *Talanta*, *81*(1-2), 346-354.

Mujahid, M., Latif, S., Ahmed, M., Shehzadi, W., Imran, M., Ahmad, M., & Mahmud, Z. (2022). Modified matrix solid phase dispersion-HPLC method for determination of pesticide residue in vegetables and their impact on human health: A risk assessment. *Frontiers in Chemistry*, *10*, 1084350.

Mukherjee, A., Goyal, T. M., Miglani, S., & Kapoor, A. (2019). SPS barriers to India's agriculture export learning from the EU experiences in SPS and food safety standards.

Navarro, P., Pérez, A., Gabaldón, J. A., Núñez-Delicado, E., Puchades, R., Maquieira, A., & Morais, S. (2013). Detection of chemical residues in tangerine juices by a duplex immunoassay. *Talanta*, 116, 33–38.

Navarro, S., Barba, A., Navarro, G., Vela, N., & Oliva, J. (2000). Multiresidue method for the rapid determination–in grape, must and wine–of fungicides frequently used on vineyards. *Journal of Chromatography A*, *882*(1-2), 221-229.

Ni, Y. X., Yang, H., Zhang, H. T., He, Q. A., Huang, S. Q., Qin, M. L., & Ma, Y. N. (2018). Analysis of four sulfonylurea herbicides in cereals using modified Quick, Easy, Cheap, Effective, Rugged, and Safe sample preparation method coupled with liquid chromatography-tandem mass spectrometry. *Journal of Chromatography A, 1537*, 27–34.

Oellig, C., & Schwack, W. (2011). Planar solid phase extraction—A new cleanup concept in multi-residue analysis of pesticides by liquid chromatography–mass spectrometry. *Journal of Chromatography A, 1218*(37), 6540-6547.

Pang, G. F., Cao, Y. Z., Zhang, J. J., Fan, C. L., Liu, Y. M., Li, X. M., ... & Guo, T. T. (2006). Validation study on 660 pesticide residues in animal tissues by gel permeation chromatography cleanup/gas chromatography–mass spectrometry and liquid chromatography–tandem mass spectrometry. *Journal of Chromatography A, 1125*(1), 1-30.

Rocca, L. M., Cecca, J., Episcopo, N., & Fabrizi, G. (2017). Ambient mass spectrometry: Direct analysis of dimethoate, tebuconazole, and trifloxystrobin on olive and vine leaves by desorption electrospray ionization interface. *Biol. Mass Spectrom, 52*, 709–719.

Saadati, N., Abdullah, M. P., Zakaria, Z., Sany, S. B. T., Rezayi, M., & Hassonizadeh, H. (2013). Limit of detection and limit of quantification development procedures for organochlorine pesticides analysis in water and sediment matrices. *Chemistry Central Journal, 7*, 1-10.

Saini, M. K., Mishra, S., Alam, S., Thakur, L. K., Singh, O., & Raza, S. K. (2016). Method development and validation of multiclass pesticide residues and metabolites in wheat by GC-ECD and GC-MS. *Asian J. Research Chem, 9*(1), 711-715.

Sanghi, R., & Kannamkumarath, S. S. (2004). Comparison of extraction methods by soxhlet, sonicator, and microwave in the screening of pesticide residues from solid matrices. *Journal of Analytical Chemistry, 59*, 1032-1036.

Sawant, C. G., Guru, P. N., Mundhe, S. G., Patil, C. S., & Sadhukhan, R. (2022). Method validation and dissipation behavior of ethion 50 EC insecticide in winter cabbage.

Saravi, S. S. S., & Shokrzadeh, M. (2011). Role of pesticides in human life in the modern age: a review. *Pesticides in the modern world-risks and benefits, 5*, 3-12.

Schierow, L. J., & Esworthy, R. (2008). *Pesticide law: a summary of the statutes.* Congressional Research Service.

Sen, R., Escorihuela, J., Smulders, M. M. J., & Zuilhof, H. (2016). Use of ambient ionization high-resolution mass spectrometry for the kinetic analysis of organic surface reactions. *Langmuir: The ACS Journal of Surfaces and Colloids, 32*(14), 3412–3419.

Syrgabek, Y., & Alimzhanova, M. (2022). Modern Analytical Methods for the Analysis of Pesticides in Grapes: A Review. *Foods, 11*(11), 1623.https://doi.org/10.3390/foods 11111623.

Taylor, M. J., Hunter, K., Hunter, K. B., Lindsay, D., & Le Bouhellec, S. (2002). Multi-residue method for rapid screening and confirmation of pesticides in crude extracts of fruits and vegetables using isocratic liquid chromatography with electrospray tandem mass spectrometry. *Journal of Chromatography A, 982*(2), 225-236.

Timofeeva, I.; Shishov, A.; Kanashina, D.; Dzema, D.; Bulatov.(2017). A. Online in-syringe sugaring-out liquid-liquid extraction coupled with HPLC-MS/MS for the determination of pesticides in fruit and berry juices. *Talanta, 167,* 761–767.

Verma, M. (2010). Regulatory Acceptance for Accelerated Solvent Extraction: A Reference Summary.http://tools.thermofisher.com/content/sfs/brochures/WP-71259-ASE-Reference-WP71259-EN.pdf.

Vilela, A. D. O., Faroni, L. R. D. A., Rodrigues, A. A. Z., Heleno, F. F., de Queiroz, M. E. L. R., Moura, E. D. S., & Gomes, J. L. (2020). Headspace Solid-Phase Microextraction: Validation of the Method and Determination of Allyl Isothiocyanate Persistence in Cowpea Beans. *ACS omega*, *5*(34), 21364-21373.

Wu, G., Bao, X., Zhao, S., Wu, J., Han, A., & Ye, Q. (2011). Analysis of multi-pesticide residues in the foods of animal origin by GC–MS coupled with accelerated solvent extraction and gel permeation chromatography cleanup. *Food Chemistry*, *126*(2), 646-654.

Yang, Q., Chen, B. B., He, M., & Hu, B. (2018). Sensitive determination of seven triazine herbicide in honey, tomato and environmental water samples by hollow fiber based liquid-liquid microextraction combined with sweeping micellar electrokinetic capillary chromatography. *Talanta*, 186, 88–96.

Yao, S., Zhao, Z., Lu, W., Dong, X., Hu, J., & Liu, X. (2021). Evaluation of dissipation behavior, residues, and dietary risk assessment of fludioxonil in cherry via QuEChERS using HPLC-MS/MS technique. *Molecules*, *26*(11), 3344.

Zhang, L., Gionfriddo, E., Acquaro, V., & Pawliszyn, J. (2018). Direct immersion solidphase microextraction analysis of multiclass contaminants in edible seaweeds by gas chromatography-mass spectrometry. *Analytica Chimica Acta, 1031*, 83–97.

Zhao, P. F., Wang, Z. K., Li, K. J., Guo, X. J., & Zhao, L. S. (2018). Multi-residue enantio meric analysis of 18 chiral pesticides in water, soil and river sediment using magnetic solid-phase extraction based on amino modified multiwalled carbon nanotubes and chiral liquid chromatography coupled with tandem mass spectrometry. *Journal of Chromatography A*, *1568*, 8–21.

Żwir-Ferenc, A., & Biziuk, M. (2004). An analysis of pesticides and polychlorinated biphenyls in biological samples and foods. *Critical reviews in analytical chemistry*, *34*(2), 95-103.

Chapter 5

Insecticide Remediation: Need, Advanced Techniques, and Future Prospects

Rakesh Kumar[1,*,$]
Rajnish Rai[2]
Vijay Kumar[3]
Mamta Kumari[1,4]
Vivek Adhikari[5]
and Satdev Rokana[1,6]

[1]Department of Soil Science, Bihar Agricultural University, Bhagalpur, India
[$] Dept. of Soil Science, GAC Jogiwala, SKRAU Bikaner – 334 006, India
[2]Department of Agricultural Entomology, GBPUA&T, Pantnagar, India
[3]Department of Entomology, RCA MPUAT Udaipur, India
[4]Dept of Soil Science, BHU, Varanasi – 221 005, India
[5]Department of Agricultural Entomology, B A College of Agriculture, AAU, Anand, India
[6] Dept of Soil Science, Faculty of Agriculture, Tantiya University, Sri Ganganagar – 335 002, India

Abstract

The excessive and indiscriminate use of insecticides has led to significant environmental pollution and adverse effects on human health, contaminating soil, water bodies, and food chains. This chapter highlights the urgent need for effective insecticide remediation to mitigate these risks. It explores advanced techniques for insecticide removal, including physical, chemical, biological, and nanotechnological

---

* Corresponding Author's Email: drrakeshmothsara@gmail.com

In: Insecticides
Editors: Archana Anokhe and Govindaraj Kamalam Dinesh
ISBN: 979-8-90134-022-6

approaches. Physical techniques, such as soil vapor extraction, activated carbon adsorption, and membrane filtration, effectively remove insecticides from contaminated soils and water. Chemical methods, such as advanced oxidation processes (AOPs) and photocatalysis, are examined for their potential to degrade insecticides into non-toxic byproducts. The chapter also emphasizes bioremediation, including microbial degradation and phytoremediation, as promising solutions for environmental remediation. Emerging technologies, particularly nanotechnology-based approaches like nanosensors and nano-remediation, are discussed for their potential to enhance the detection and removal of insecticides at low concentrations. The chapter concludes by stressing the importance of interdisciplinary research collaborations and policy interventions. It advocates for increased awareness, education, and the adoption of sustainable pest management practices to reduce reliance on insecticides and promote environmentally friendly alternatives.

Keywords: insecticide pollution, insecticide remediation, nanotechnology, maximum residual limit (MRL)

Introduction

In today's modern environmental and agricultural landscapes, the issue of insecticide contamination has reached alarming levels. The excessive and indiscriminate use of insecticides has resulted in significant environmental pollution and adverse effects on human health. Insecticide contamination in soil, water bodies, and food chains has become a growing concern worldwide, posing a threat to environmental sustainability, agricultural practices, and ecosystem health.

The urgent need for advanced techniques in addressing this critical issue of insecticide contamination cannot be overstated. It is crucial to develop effective strategies and technologies for removing and degrading insecticides to mitigate the environmental and health risks associated with their presence. This chapter delves into the realm of insecticide remediation, exploring advanced techniques that offer hope in tackling this pressing challenge. The chapter begins by emphasizing the importance of insecticide remediation in mitigating the environmental and health risks posed by insecticide contamination. It highlights the need for effective strategies and technologies to address this issue and restore balance to our ecosystems. The chapter then proceeds to explore a range of advanced techniques for insecticide removal and degradation, including physical, chemical, biological, and nanotechnological approaches.

By exploring physical techniques such as soil vapor extraction, activated carbon adsorption, and membrane filtration, the chapter demonstrates practical methods for removing insecticides from contaminated soils and water. It explores chemical methods, such as advanced oxidation processes (AOPs) and photocatalysis, which hold promise in degrading insecticides into non-toxic byproducts. The chapter also highlights the role of bioremediation, including microbial degradation and phytoremediation, in addressing insecticide contamination.

Furthermore, the chapter sheds light on emerging technologies and prospects in insecticide remediation. It explores the potential of nanotechnology-based approaches, such as nanosensors and nanoremediation, in enhancing the detection and removal of insecticides at low concentrations. These advancements offer exciting opportunities for precise monitoring and efficient remediation of insecticide contamination.

This chapter highlights the importance of insecticide remediation in contemporary environmental and agricultural landscapes. It highlights the urgent need for advanced techniques to address this critical issue and restore ecosystem health. By adopting sustainable pest management practices and promoting environmentally friendly alternatives, we can work towards a future where insecticide contamination is effectively mitigated. This chapter provides a roadmap for the insightful exploration that lies ahead, offering valuable insights and guidance for researchers, policymakers, and stakeholders as they work to achieve a healthier and more sustainable world.

Advanced Techniques for Insecticide Remediation

In the quest for viable solutions to insecticide contamination, a spectrum of advanced techniques emerges as beacons of hope. These techniques can be classified into physical, chemical, biological, and nanotechnological approaches. Figure 1 represents the classification of insecticide remediation techniques.

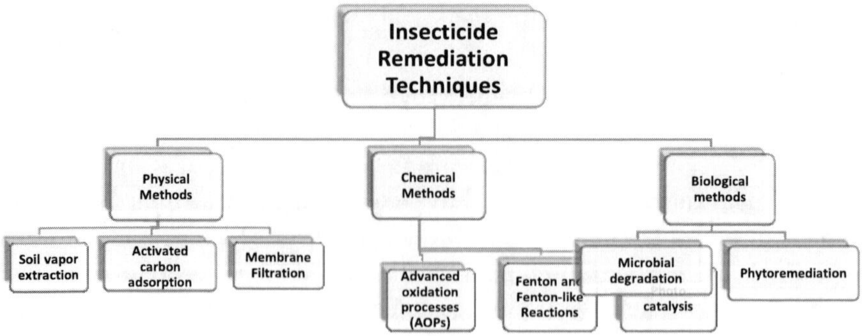

Figure 1. Classification of Insecticide Remediation Techniques.

Physical Techniques

Physical techniques focus on removing insecticides from contaminated matrices, with methods such as soil vapor extraction, activated carbon adsorption, and membrane filtration taking center stage. Soil vapour extraction (SVE) stands as a prevalent and generally successful physical remediation method for addressing volatile pollutants present in vadose zone (unsaturated) soils. The SVE process involves extracting contaminated soil gas from the vadose zone, treating it, and eventually releasing it into the atmosphere (Anderson et al., 2001). This technique proves particularly advantageous in areas where air movement can be induced through polluted zones. Over time, the efficacy of SVE may decline, attributed to the gradual removal of easily extractable contaminant mass, eventually leading to the predominance of mass transfer limitations dictating the retrieval of the residual contaminant mass. This approach has achieved a mature status and has been implemented in various configurations (SVE, 2022). Activated carbon adsorption utilizes the remarkable adsorptive properties of activated carbon to bind and concentrate insecticides, thereby decontaminating water and soil matrices (Ahmed et al., 2014; Inyang et al., 2016). Chang et al., (2014) employed a two-stage process using potassium hydroxide (KOH) to transform waste biomass (rice straw) into activated carbon with a high surface area and micropores. The resulting activated carbon was then evaluated for its capacity to remove carbofuran from an aqueous solution. The study demonstrated that rice straw can be a very effective adsorbent for carbofuran, with an adsorption capacity of 296.52 mg/g. Different adsorption parameters, including the initial carbofuran concentration, contact time, temperature, and pH, were studied for carbofuran adsorption using a batch system. This finding suggests that activated carbon prepared from rice straw could be a promising new material for the remediation of carbofuran-contaminated water bodies. Zahoor (2010) explored the elimination of thiram from aqueous solutions using activated carbon.

The adsorption of insecticides onto activated carbon is influenced by several factors, including the type of insecticide, its concentration, the solution pH, and the temperature. Generally, the adsorption of insecticides is more efficient at low pH levels and low temperatures. Activated carbon adsorption is a versatile and effective technique for insecticide remediation. It can be used to remove insecticides from water, soil, and air. Activated carbon adsorption is also relatively inexpensive and easy to implement.

Benefits of using activated carbon adsorption for insecticide remediation:

- It is a non-toxic and environmentally friendly remediation technique.
- It is effective for removing a wide range of insecticides.
- It can be used to remove insecticides from both water and soil.
- It is relatively inexpensive and easy to implement.

Limitations of using activated carbon adsorption for insecticide remediation:

- The efficiency of the adsorption process can be affected by the type of insecticide and the pH of the solution.
- The activated carbon can become saturated with insecticides and will need to be replaced periodically.
- The activated carbon can be challenging to remove from the water or soil after the remediation process is complete.

Overall, activated carbon adsorption is a promising technique for the remediation of insecticides. It is a non-toxic, environmentally friendly, and effective remediation technique that can be used to remove insecticides from water and soil.

Meanwhile, membrane filtration exploits selectively permeable membranes to separate insecticide-laden water into purified streams. Goh et al., (2022) reviewed remedial strategies for pesticides in an aqueous environment using membrane-based processes. The potentials and applications of three prevailing membrane processes, namely, NF (nanofiltration), RO (reverse osmosis), and FO (forward osmosis) for treating pesticide-containing wastewater were discussed in terms of the development of advanced membranes, separation mechanisms, and system design.

Nanofiltration (NF) and reverse osmosis (RO) are two pressure-driven membrane processes commonly used for removing micropollutants, including pesticides. RO has demonstrated a high removal efficiency of >90% for a broad spectrum of trace organic compounds (TrOCs), including those with a low molecular weight of <200 g/mol (Shin et al., 2022). NF has a lower pore size than RO, which allows it to remove smaller molecules, such as pesticides. However, NF is less effective at removing larger molecules, such as NMO (Nano Metal oxides). The efficacy of nanofilters can vary depending on the properties of the compounds with which they interact. For example, the increased surface negativity of the membrane surface upon NOM adsorption can facilitate the passage of diuron, which has a higher dipole moment despite its large molecular size (Musbah et al., 2018). Overall, NF and RO are both practical techniques for removing micropollutants, including pesticides. RO is generally more effective than NF, but it is also more expensive. The choice of which technique to use will depend on the specific application and the budget constraints.

Forward osmosis (FO) is an alternative to pressure-driven membrane filtration, utilizing a concentration gradient to drive water through a semipermeable membrane. Existing FO membranes have been ineffective in rejecting small neutral organic pollutants; however, biomimetic aquaporin membranes have demonstrated rejection values above 97% for all three trace organics tested. This is because aquaporin membranes utilize diffusion to reject pollutants, whereas cellulose acetate membranes employ steric hindrance to achieve this. FO using biomimetic aquaporin membranes is a promising technique for the remediation of water contaminated with small neutral organic pollutants. (Madsen et al., 2015).

Advantages of forward osmosis (FO) using biomimetic aquaporin membranes (Couto et al., 2020; Tibi et al., 2020):

- Low energy consumption: FO utilizes a concentration gradient to drive water through the membrane, which is less energy-intensive than pressure-driven processes, such as reverse osmosis (RO).
- Lower fouling tendency: Biomimetic aquaporin membranes exhibit a lower fouling tendency than traditional FO membranes, making them more reliable and cost-effective over time.

- Effective removal of small neutral organic pollutants: Biomimetic aquaporin membranes have been demonstrated to be effective in removing small neutral organic pollutants, which are often challenging to remove using conventional methods.
- Non-toxic: FO using biomimetic aquaporin membranes is a non-toxic remediation technique, which makes it safe for use in drinking water and other applications.
- Scalable: FO using biomimetic aquaporin membranes can be scaled up to treat large volumes of water, making it a viable option for industrial and municipal water treatment.

Chemical Methods

Chemical techniques embrace the transformation of insecticides into benign byproducts. Advanced oxidation processes (AOPs), Fenton and Fenton-like Reactions, and photocatalysis stand out in this arena.

Advanced oxidation processes (AOPs) are a group of water treatment methods that utilize highly reactive species, such as hydroxyl radicals, to degrade organic pollutants, including pesticides, in wastewater. AOPs employ powerful oxidants to break down insecticides into simpler compounds, which are subsequently mineralized into innocuous molecules. AOPs can be classified into different types based on the source of the oxidizing agents, such as ozone, hydrogen peroxide, persulfate, or photocatalysts. Some of the most common AOPs for pesticidal removal are:

- Ozone-based AOPs: These processes use ozone alone or in combination with ultraviolet (UV) radiation or catalysts to generate hydroxyl radicals. Ozone-based AOPs can achieve high removal efficiencies for some pesticides, such as deltamethrin.
- Photocatalytic AOPs: These processes use UV radiation and a semiconductor catalyst, such as titanium dioxide ($TiO2$), to produce hydroxyl radicals on the catalyst surface. Photocatalytic AOPs can enhance the degradation of some pesticides, such as acetamiprid, compared to UV alone.
- Fenton-based AOPs: These processes use hydrogen peroxide and a ferrous ion catalyst to produce hydroxyl radicals in acidic conditions. Fenton-based AOPs can be combined with UV radiation to increase

the production of hydroxyl radicals and improve the removal of some pesticides, such as chlorpyrifos.

- Persulfate-based AOPs: These processes utilize persulfate, a strong oxidant, either alone or in combination with UV radiation or a catalyst, such as iron, to generate sulfate radicals. Persulfate-based AOPs can also effectively degrade some pesticides, such as acetamiprid.

AOPs have several advantages over conventional biological treatment methods, including the ability to degrade recalcitrant compounds, the absence of toxic byproducts, and the potential for achieving complete mineralization. However, AOPs also have some limitations, such as high energy consumption, low selectivity, and high operational costs. Therefore, AOPs are usually applied as tertiary treatment methods after biological processes to achieve higher water quality standards (Pham et al., 2020). Andreozzi et al., (1999) investigated the degradation of 2-chloro-4,6-diamino-1,3,5-triazine in aqueous solutions using the photo-Fenton process. Photocatalysis relies on catalysts activated by light to drive chemical reactions that degrade insecticides. These techniques hold immense promise in dismantling complex insecticide molecules into less harmful constituents. Khataee and Kasiri (2010) reviewed the photocatalytic degradation of organic dyes containing various functional groups using TiO_2 nanomaterials in an aqueous solution. They have discussed the photocatalytic properties of nanostructured TiO_2 and concluded that the photocatalytic degradation rate strongly depends on the basic structure of the molecule and the nature of auxiliary groups attached to the aromatic nuclei of the dyes. The influences of various substitutes, including alkyl side chains, methyl, nitrate, hydroxyl, and carboxylic groups, as well as the presence of a chloro atom, have been discussed in detail.

Biological Approaches

These approaches extol the roles of bioremediation in addressing insecticide contamination. Microbial degradation harnesses the metabolic prowess of microorganisms to transform insecticides into non-toxic metabolites through biological processes. An array of microbes, including species of *Bacillus*, *Klebsiella*, *Mycobacterium*, *Pandoraea*, *Pseudomonas*, *Sphingomonas*, *Sphingopyxis*, *Rhodococcus*, *Sphingobium*, and *Phanerochaete chrysosporium*, reflects their potential in bioremediating a wide range of pesticides, ultimately revitalizing soil quality (Dai et al., 2004). Microbial degradation is now the most comprehensive and widespread form of biodegradation. Bacteria, fungi, and algae are all used in various techniques, with bacteria being the most common. A considerable number of bacteria that degrade chlorpyrifos have been reported and isolated to date. Huang et al., (2021) reviewed more than 38 bacterial, 10 fungal, and 2 algal species from the genera *Alcaligenes*, *Diaphorobacter*, *Pseudomonas*, *Shewanella*, *Stenotrophomonas*, *Xanthomonas*, *Rhizobium*, *Ochrobactrum*, *Paracoccus*, *Gordonia*, and *Cupriavidus*, which are capable of degrading chlorpyrifos. Singh et al., (2011) isolated a cyanobacterium, *Synechocystis sp.* strain PUPCCC 64, from rice fields. This cyanobacterium was able to eliminate 93.8% of chlorpyrifos (5 mg/L) within eight days in a medium. Cyanobacteria are a relatively inexpensive and easy-to-grow organism, making them a viable option for large-scale remediation projects. In paddy-growing developing Asian countries, where chlorpyrifos is a widely used pesticide, this study could have a significant impact on the remediation of water contaminated with this toxic substance. Microbes that degrade a wide range of insecticides, including organophosphates and carbamates, have been discovered. Microorganisms play a pivotal role in the degradation of carbamate compounds through intricate biochemical pathways. The primary modes of microbial degradation of insecticides are hydrolysis, oxidation, reduction, and use as a carbon source (Mishra et al., 2021).

Phytoremediation, on the other hand, involves using plants to extract and accumulate pollutants, such as insecticides, from soil and water. It is a potential strategy for cleaning up heavy metal-polluted soil. It has been utilized to reduce contaminants such as metals, herbicides, solvents, explosives, crude oil, and its derivatives. Phytoremediation is a low-cost, environmentally friendly method of cleaning up contaminated soil, air, and water. These compounds are absorbed by plants from the soil and metabolized into simpler forms through various processes, including phytovolatilization, phytostimulation, phytoextraction, phytofiltration, rhizofiltration, and rhizodegradation (Takkar et al., 2022).

Species such as the Blue Flag iris (*Iris versicolor*), broom sedge (*Andropogon virginicus*), and switchgrass (*Panicum virgatum*) are commonly employed in this approach for the remediation of pesticides and heavy metals from soil. The roots of these plants exhibit a remarkable ability to absorb and sequester insecticides, thereby alleviating contamination and promoting soil recovery. Poplar trees have been successfully used in the phytodegradation of hazardous and recalcitrant chemical substances. Poplar and willow trees have also demonstrated a substantial potential to absorb and translocate/degrade chlorpyrifos. Furthermore, transgenic poplar plants expressing the rabbit pon1 gene, which is known to be involved in organophosphate (OP) pesticide degradation pathways, were created and tested for enhanced resistance and chlorpyrifos elimination (Lee, K. Y., 2014). Phytoremediation offers several advantages over traditional remediation techniques. Because it utilizes plants and natural resources, it is often less expensive. Remediation is performed on-site, saving money on transportation and off-site processing. Plant roots stabilize soil and prevent contaminants from moving through runoff and wind-blown dust. Furthermore, phytoremediation maintains topsoil, preserving soil fertility.

Assessment of wetland plants such as *Cyperus alternifolius* and *Cyperus papyrus* by Liu et al., (2021) revealed that neonicotinoids like imidacloprid, thiacloprid, and acetamiprid were readily removed by biodegradation and plant accumulation.

Emerging Technologies in the Field of Nano-Material-Based Remediation

As the boundaries of science continue to expand, new horizons beckon in the field of insecticide remediation. The emergence of nanotechnology heralds exciting opportunities for detecting and eradicating insecticides at exceedingly low concentrations.

Nanosensors, capable of detecting minuscule amounts of insecticides, offer a precise and efficient means of monitoring contamination levels. Types of nanosensors for pesticide detection and their mechanism (Zia et al., 2022):

Carbon-Based Nanosensors

These sensors are composed of nanomaterials, a specialized identification element, and a signal transduction mechanism. Due to their high electrical conductivity, large surface area, high thermal conductivity, and mechanical strength, carbon-based nanomaterials such as nanotubes (CNTs), carbon nanofibers, and graphene are frequently utilized in nano-enabled sensors. Due to their superior electronic properties, graphene-based sensors have been utilized for fluorescence quenching. Carbon nanotubes are composed of rolled graphene sheets, which have cylindrical shapes with diameters measured in several nanometers. Carbon nanotubes can vary in length, diameter, size, and number of layers. They are divided into two types based on their structure: single-walled nanotubes (SWCNT) and multi-walled nanotubes (MWNT).

Metallic Nanoparticles-Based Nanosensors

Metallic nanoparticles, such as silver, copper, zinc oxide, gold, titanium dioxide (TiO2), copper oxide, and iron oxide, have been utilized for pesticide detection. Magnetic iron oxides, including magnetite (Fe_3O_4) and maghemite (γ-Fe_3O_4), can be easily functionalized with antibodies, ligands, and other capping agents for pesticide detection. TiO_2 is also used in the design of nanosensors due to its outstanding physical and chemical properties.

Quantum Dots (QDs) Based Nanosensors

QDs have been employed as probes for environmental contaminants such as pesticides, heavy metals, and other poisons. The sensitivity of QD luminescent behaviour to the surface state of nanomaterials determines detection. Sensing responses result from physical or chemical interactions with analyte molecules, which either inhibit or enhance photoluminescence. The primary sensing mechanism of QD sensing assemblies is due to the transfer of energy between target analyte molecules and the QD's fluorophore, rather than direct contact with the QD surface.

A MnO_2 nanosheet, QDs modified, FRET (Fluorescence Resonance Energy Transfer)-based sensor has been developed for the detection of organophosphates (OPs). QDs are used as fluorescence units, while MnO_2 is employed as a dual-functional nano quencher and receptor for the development of the sensing platform. The sensing mechanism is based on the butyrylcholinesterase (BChE)-induced decomposition of MnO_2 nanosheets. When OPs are present, they inhibit BChE, which prevents the decomposition of MnO_2 nanosheets. This results in an increase in the fluorescence intensity of QDs, which can be used to detect the presence of OPs.

Molecularly Imprinted Polymers (MIPs) Based Nanosensors

MIPs are precise synthetic receptors that, like antigen-antibody interactions, provide structure-based recognition active sites for the attachment of desired analyte molecules. However, MIPs are more suited for biosensing applications due to their exceptional chemical/thermal stability, which results in an extended shelf life. MIPs are used to selectively remove desirable molecules from interfering compounds.

The mechanism of these sensors involves the recognition of the pesticide molecules by the specific recognition element present in the sensor. This recognition leads to a change in the signal transduction process, which can be detected and measured. For example, in the case of QDs, the sensing response results from the interactions, either physical or chemical, with analyte molecules, leading to quenching or improvement of photoluminescence.

Boregowda et al., (2022) reviewed recent advances in nanoremediation. They identified 16 nanomaterials that can be utilized as nanosensors for detecting pesticide residues. Figure 2 represents a diagram of nanosensors detecting pollutants.

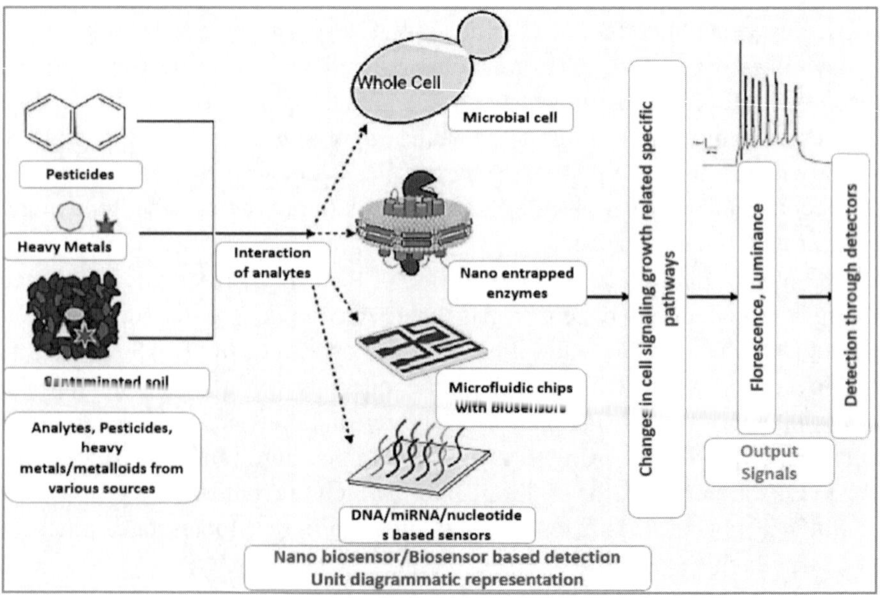

Figure 2. Diagram representing Nanosensors detecting pollutants.

Additionally, nanoremediation involves the use of nanoparticles to catalyze the breakdown of insecticides. Nanosilver, titanium dioxide, and zinc oxide are commonly used as photocatalysts for the breakdown of pesticides. Heterogeneous photocatalysis, which utilizes light energy and a semiconductor, has been widely employed for the cleanup of organic pollution. Titania (TiO_2)-based materials supported on zeolitic matrices were synthesized through in situ TiO_2 production to degrade DDVP (Dichlorvos) (Sengupta, S., & Bhattacharya, 2019).

Nano-Bioremediation (NBR)

Combining nanoremediation with bioremediation has shown a more feasible application, giving rise to the domain of Nano-bioremediation (NBR). NBR enhances the bioremediation process by using nanoparticles generated from prokaryotes (Gram-negative rods, actinobacteria) and eukaryotes (fungi, algae, and plants). A combination of nano zero-valent iron (nZVI), surfactants, and electrokinetic treatments has also been shown to be effective for remediation of heavy metal, polychlorinated biphenyl (PCB), pesticide, and chlorinated volatile organic compound (cVOC)- contaminated sites (Bhattacharya et al., 2016). NBR also involves the immobilization of entire cells as well as the use of nanoparticles. Recombinant *Escherichia coli* cells expressing the long-lasting atrazine-dechlorinating enzyme were enclosed in a porous gel composed of siliconized silica nanoparticles (SiNPs). The encapsulated cells degraded atrazine three times faster than loose cells (Reátegui et al., 2012). Zhuang et al., (2015) provided another example of NBR employing cell immobilization. To decompose the heterocyclic aromatic chemical molecule quinoline, they employed Streptomyces sp. N01 immobilized on bamboo-carbon supported Fe3O4 nanoparticles (Fe3O4/BC).

Nano-Biosurfactants

Biosurfactants are amphiphilic compounds produced on the surfaces of microbial cells or released extracellularly and can self-assemble into micelles. In the organic phase, they can also self-assemble into reverse micelles. These can self-assemble to form nano-sized micelles in the aqueous phase, and reverse micelles in the organic phase are known as nano-biosurfactants. Rhamnolipids produced by Pseudomonas aeruginosa and glycolipids and sophorose lipids produced by bacteria for the degradation of polycyclic aromatic hydrocarbons (PAHs) are two examples of nano-biosurfactants utilized in pesticide degradation (Boregowda et al., 2022). These technologies have the potential to revolutionize our ability to tackle contamination at its source, paving the way for more proactive and practical solutions. Figure 3 represents the diagrammatic representation of biosurfactants.

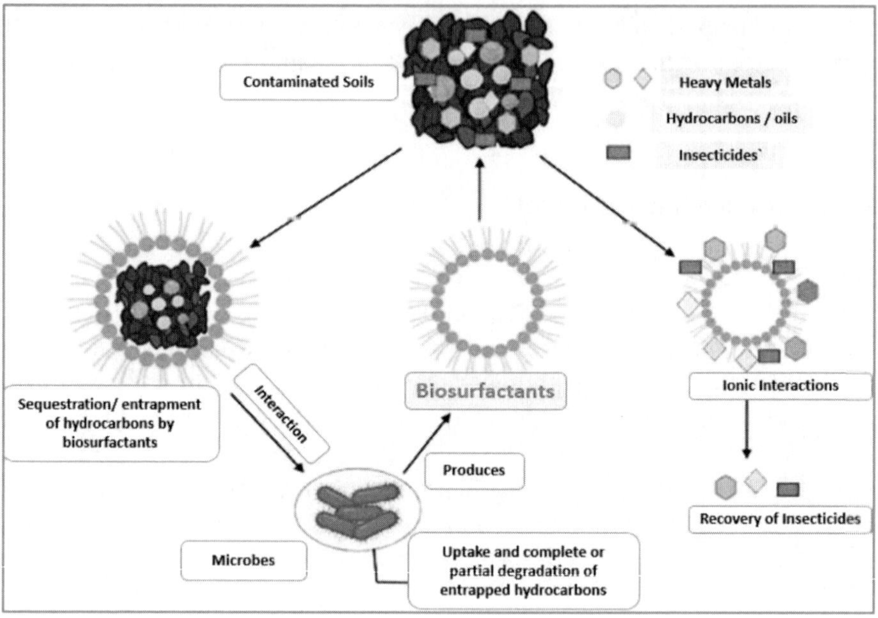

Figure 3. Diagrammatic representation of biosurfactants.

Collaborations and Policy Interventions

In reflecting on the journey of insecticide remediation, the chapter underscores the critical importance of collaborative efforts and policy interventions in addressing the multifaceted challenges posed by insecticide contamination. The complexities of insecticide pollution transcend disciplinary boundaries, necessitating the collective wisdom and expertise of scientists, engineers, ecologists, policymakers, and the broader community. Insecticide remediation policies aim to regulate the entire lifecycle of pesticides, from manufacturing to disposal, to ensure their safe and judicious use, minimize risks to humans, animals, and the environment, and promote sustainable agriculture. These policies consist of several components that work together to achieve their objectives. Here are the key components and examples of how they work (NAAS, 2020):

- Regulation and Legislation: The development and implementation of laws and regulations to govern the use of pesticides. For example, India approved a new Pesticide Management Bill in 2020 to regulate the manufacture, import, sale, storage, distribution, use, and disposal of pesticides.
- Quality Control: Ensuring that pesticides meet safety standards and are effective in controlling pests, for example, setting maximum residue limits (MRLs) for pesticide residues on agricultural commodities and implementing stringent quality testing procedures.
- Education and Awareness: Promoting education and awareness programs to inform farmers and other stakeholders about the safe and judicious use of pesticides. For example, training programs, workshops, and campaigns to raise awareness about the potential risks associated with pesticide use and the importance of following recommended practices
- Monitoring and Enforcement: Establishing monitoring systems to track the use of pesticides and enforce compliance with regulations. For example, regular inspections, audits, and penalties for non-compliance. It may also include the establishment of dedicated regulatory bodies or agencies responsible for overseeing pesticide management.
- Research and Development: Investing in research and development to develop alternative pest management strategies that reduce reliance on chemical insecticides. For example, the development of integrated pest management (IPM) approaches, biological control methods, and the use of resistant crop varieties.
- International Collaboration: Collaborating with international organizations and other countries to share best practices, exchange information, and harmonize pesticide regulations. For example, the Food and Agriculture Organization (FAO) of the United Nations collaborates with countries to develop and implement policies and programs for pesticide management.

By implementing these components, insecticide remediation policies aim to minimize the risks associated with pesticide use, protect human health and the environment, and promote sustainable agricultural practices.

Interdisciplinary Collaborations

Interdisciplinary research is the cornerstone upon which innovative solutions are built. The chapter stresses that the convergence of diverse perspectives enhances problem-solving, leading to novel insights and breakthrough discoveries. Scientists can collaborate with engineers to design and implement efficient remediation systems. Ecologists contribute their understanding of ecosystem dynamics, highlighting potential ripple effects caused by insecticide pollution. Policymakers bridge the gap between science and society, translating research findings into actionable policies that promote sustainable pest management practices.

Furthermore, the chapter emphasizes the importance of fostering a culture of interdisciplinary collaboration from early education through advanced research. Encouraging scientists to step outside their comfort zones and collaborate across traditional boundaries not only fuels innovation but also cultivates a holistic approach to problem-solving that mirrors the intricate web of ecosystems.

Policy Interventions for a Sustainable Future

Insecticide contamination demands a multifaceted response that extends beyond scientific innovation. Effective policy interventions play a pivotal role in regulating insecticide use, encouraging sustainable alternatives, and mitigating the environmental and health risks associated with these chemicals.

Regulation is a crucial instrument in controlling the excessive use of insecticides. Stringent limits on usage, especially for highly toxic compounds, can prevent overreliance and curb contamination. Additionally, policies can incentivize the adoption of integrated pest management (IPM) strategies that emphasize a holistic approach to pest control, encompassing biological, cultural, and chemical methods in harmony with the environment.

The chapter underscores the significance of creating a conducive environment for innovation. Governments can fund research initiatives aimed at developing advanced remediation technologies and sustainable alternatives to traditional insecticides. Tax incentives and subsidies for adopting environmentally friendly practices can encourage industry and agriculture to adopt responsible pest management approaches.

Navigating Future Prospects

As the chapter draws to a close, it ventures into the exciting realm of prospects in insecticide remediation. The dynamic nature of science and technology promises a trajectory of possibilities once unimaginable. Nanotechnology, a discipline that manipulates matter at the nanoscale, holds significant potential in combating insecticide contamination. Nanosensors, capable of detecting infinitesimal amounts of insecticides, offer a sensitive and efficient method for monitoring contamination levels. Their applications extend beyond traditional monitoring, enabling real-time data collection and analysis to inform decision-making. Nanoremediation, an emerging concept, harnesses the catalytic properties of nanoparticles. These miniature agents can accelerate the degradation of insecticides, facilitating their conversion into harmless byproducts. While the technology is still in its infancy, the chapter highlights its immense promise in revolutionizing the field of remediation.

Embracing Green Chemistry

An underlying theme that resonates throughout the chapter is the need to adopt more sustainable practices. Green chemistry, a philosophy that champions the design of chemical products and processes that minimize environmental impact, aligns seamlessly with the objectives of insecticide remediation. The future of pest management lies in the development of insecticides that degrade rapidly and do not persist in the environment, thus reducing the risk of contamination.

Conclusion: Paving the Path Forward

In the realm of insecticide contamination, the past is fraught with missteps, the present riddled with challenges, and the future laden with potential. The chapter navigated through the intricate tapestry of insecticide remediation, unveiling the urgency of addressing environmental pollution and health risks. Advanced techniques, ranging from physical and chemical methods to biological and nanotechnological approaches, underscore our capacity to engineer transformative solutions.

The key contributions and insights can be summarized as follows:

- Environmental and Agricultural Significance: The chapter emphasizes the pressing need for insecticide remediation in contemporary environmental and agricultural landscapes. It emphasizes the detrimental effects of insecticide contamination on environmental sustainability, agricultural practices, and ecosystem health. The chapter underscores the importance of addressing this critical issue to safeguard our ecosystems and promote sustainable pest management.
- Advanced Techniques: The chapter explores a range of advanced techniques for insecticide remediation, including physical, chemical, biological, and nanotechnological approaches. It provides detailed insights into each technique, including soil vapor extraction, activated carbon adsorption, membrane filtration, advanced oxidation processes (AOPs), photocatalysis, microbial degradation, phytoremediation, nanosensors, and nanoremediation. These techniques offer promising solutions for the removal and degradation of insecticides, enabling effective remediation of contaminated soils, water bodies, and air.
- Reshaping Environmental and Agricultural Practices: The advanced techniques discussed in the chapter have the potential to reshape environmental and agricultural practices. By adopting these techniques, we can reduce the reliance on harmful insecticides and promote environmentally friendly alternatives. The chapter emphasizes the importance of interdisciplinary research collaborations and policy interventions in driving the adoption of these advanced techniques and promoting sustainable pest management practices.
- Safeguarding Ecosystem Health: This chapter examines the broader implications of advanced insecticide remediation techniques for safeguarding ecosystem health. By effectively removing and degrading insecticides, we can minimize their adverse effects on ecosystems and protect biodiversity. The chapter highlights the role

of bioremediation, nanotechnology, and policy interventions in achieving this goal.

In conclusion, the chapter on insecticide remediation provides a comprehensive understanding of the significance of advanced techniques in addressing insecticide contamination. It highlights the potential of these techniques to transform environmental and agricultural practices, foster sustainable pest management, and protect ecosystem health. By adopting these advanced techniques and implementing appropriate policies, we can work towards a future where insecticide contamination is effectively mitigated, and our ecosystems thrive more healthily and sustainably.

References

Ahmad, M., Rajapaksha, A. U., Lim, J. E., Zhang, M., Bolan, N., Mohan, D., ... & Ok, Y. S. (2014). Biochar as a sorbent for contaminant management in soil and water: A review. *Chemosphere*, 99, 19-33.

Anderson, R. H., & Clay, N. (2001). Soil vapor extraction—A proven technology. *Ground Water Monitoring & Remediation*, 21(4), 92-98.

Andreozzi, R., Caprio, V., & Insola, A. (1999). Advanced oxidation processes (AOP) for water purification and recovery. *Catalysis Today*, 53(1), 51-59.

Bhattacharya, K., Mukherjee, S. P., Gallud, A., Burkert, S. C., Bistarelli, S., Bellucci, S., ... & Fadeel, B. (2016). Biological interactions of carbon-based nanomaterials: From coronation to degradation. *Nanomedicine: Nanotechnology, Biology and Medicine*, 12(2), 333-351.

Boregowda, N., Jogigowda, S. C., Bhavya, G., Sunilkumar, C. R., Geetha, N., Udikeri, S. S., ... & Jogaiah, S. (2022). Recent advances in nanoremediation: Carving sustainable solution to clean-up polluted agriculture soils. *Environmental Pollution*, 297, 118728.

Chang, K. L., Lin, J. H., & Chen, S. T. (2011). Adsorption studies on the removal of pesticides (Carbofuran) using activated carbon from rice straw agricultural waste. *International Journal of Agricultural and Biosystems Engineering*, 5(4), 210-213.

Couto, C. F., Santos, A. V., Amaral, M. C. S., Lange, L. C., de Andrade, L. H., Foureaux, A. F. S., & Fernandes, B. S. (2020). Assessing potential of nanofiltration, reverse osmosis and membrane distillation drinking water treatment for pharmaceutically active compounds (PhACs) removal. *Journal of Water Process Engineering*, 33, 101029.

Dai M. H. et al., Genome shuffling improves degradation of the anthropogenic pesticide Pentachlorophenol by *Sphingobium chlorophenolicum* ATCC 39723 *Appl. Environ. Microbiol.* (2004).

Huang, Y., Zhang, W., Pang, S., Chen, J., Bhatt, P., Mishra, S., & Chen, S. (2021). Insights into the microbial degradation and catalytic mechanisms of chlorpyrifos. *Environmental Research*, 194, 110660.

Inyang, M., Gao, B., Yao, Y., Xue, Y., Zimmerman, A. R., Mosa, A., ... & Cao, X. (2016). Removal of heavy metals from aqueous solution by biochars derived from anaerobically digested biomass. *Bioresource Technology*, 214, 858-862.

Lee, K. Y. (2014). Phytoremediation of chlorpyrifos insecticide: the use of woody plants and transgenics to enhance and understand the uptake, translocation, and transformation of chlorpyrifos (Doctoral dissertation).

Liu, H., Tang, X., Xu, X., Dai, Y., Zhang, X., & Yang, Y. (2021). Potential for phytoremediation of neonicotinoids by nine wetland plants. *Chemosphere*, 283, 131083.

Madsen, H. T., Bajraktari, N., Hélix-Nielsen, C., Van der Bruggen, B., & Søgaard, E. G. (2015). Use of biomimetic forward osmosis membrane for trace organics removal. *Journal of Membrane Science*, 476, 469-474.

Mishra, S., Pang, S., Zhang, W., Lin, Z., Bhatt, P., & Chen, S. (2021). Insights into the microbial degradation and biochemical mechanisms of carbamates. *Chemosphere*, 279, 130500.

Musbah, I., Cicéron, D., Saboni, A., & Alexandrova, S. (2013). Retention of pesticides and metabolites by nanofiltration by effects of size and dipole moment. *Desalination*, 313, 51-56.

NAAS. (2020). Better Management of Pesticides in India: Policy Perspectives Policy Brief No. 6. *National Academy of Agricultural Sciences*, New Delhi. Retrieved from https://ppl-ai-file-upload.s3.amazonaws.com/web/direct-files/2060138/72d11a5a-f9af-4252-a891-e8d63eeba5e5/Better Management of Pesticide.pdf.

Pham, T. H., Bui, H. M., & Bui, T. X. (2020). Advanced oxidation processes for the removal of pesticides. In *Current Developments in Biotechnology and Bioengineering* (pp. 309-330). Elsevier.

Reátegui, E., Reynolds, E., Kasinkas, L., Aggarwal, A., Sadowsky, M. J., Aksan, A., & Wackett, L. P. (2012). Silica gel-encapsulated AtzA biocatalyst for atrazine biodegradation. *Applied microbiology and biotechnology*, 96, 231-240.

Sengupta, S., & Bhattacharya. 2019. Nano-Remediation: Carving solution to Pesticide Pollution, *Agrobiosnewsletter*, P. 90.

Shin, M. G., Choi, W., Park, S. J., Jeon, S., Hong, S., & Lee, J. H. (2022). Critical review and comprehensive analysis of trace organic compound (TOrC) removal with polyamide RO/NF membranes: Mechanisms and materials. *Chemical Engineering Journal*, 427, 130957.

Singh, D. P., Khattar, J. I. S., Nadda, J., Singh, Y., Garg, A., Kaur, N., & Gulati, A. (2011). Chlorpyrifos degradation by the cyanobacterium Synechocystis sp. strain PUPCCC 64. *Environmental Science and Pollution Research*, 18, 1351-1359.

Soil Vapor Extraction (SVE). (2022, February 21). *Enviro Wiki.* Retrieved 05:31, August 8, 2023 from https://www.enviro.wiki/index.php?title=Soil_Vapor_Extraction_(SVE)&oldid=15442.

Takkar, S., Shandilya, C., Agrahari, R., Chaurasia, A., Vishwakarma, K., Mohapatra, S., ... & Mishra, A. (2022). Green technology: Phytoremediation for pesticide pollution. In

Phytoremediation Technology for the Removal of Heavy Metals and Other Contaminants from Soil and Water (pp. 353-375). Elsevier.

Tibi, F., Charfi, A., Cho, J., & Kim, J. (2020). Fabrication of polymeric membranes for membrane distillation process and application for wastewater treatment: Critical review. *Process Safety and Environmental Protection*, 141, 190-201.

Zahoor, M. (2010). Removal of thiram from aqueous solutions. *Journal of the Chinese Chemical Society*, 57(6), 1361-1366.

Zhuang, H., Han, H., Xu, P., Hou, B., Jia, S., Wang, D., & Li, K. (2015). Biodegradation of quinoline by Streptomyces sp. N01 immobilized on bamboo carbon supported Fe3O4 nanoparticles. *Biochemical Engineering Journal*, 99, 44-47.

Zia, R., Taj, A., Younis, S., Bukhari, S. Z., Latif, F., Feroz, Y., ... & Bajwa, S. Z. (2022). Application of nanosensors for pesticide detection. In *Nanosensors for Smart Agriculture* (pp. 259-302). Elsevier.

Chapter 6

Impacts of Insecticides on Human Health

Sushil Kumar[1,*]
Saniya Tyagi[1,†]
and Archana Anokhe[2]
[1]Department of Entomology, B. R. D. P. G. College, Deoria, Uttar Pradesh, India
[2]Division of Entomology, ICAR-Indian Agricultural Research Institute, New Delhi, India

Abstract

Current agriculture relies heavily on pesticides to meet the growing needs of an increasingly populous world. They are the most successful group of organisms in every environmental niche and are also pests of humans and agriculture. Pesticides are widely used for their control in agriculture, households, and disease vector management, among other applications. Irrational use, however, has adverse effects on food, water, air, and soil, causing pollution and problems such as insect resistance, pest resurgence, and impacts on non-target organisms. They are toxic and pose serious risks to human health and the environment. Human diseases have been known to occur due to pesticide exposures, including Alzheimer's, amyotrophic lateral sclerosis, Parkinson's, infertility, bronchitis, asthma, congenital disabilities, autism, diabetes, obesity, respiratory disorders, organ diseases, and system failures. Also increases the risk of leukaemia, brain tumours, breast cancers, lung, stomach, liver, and urinary bladder cancers. To prevent food contamination and maintain a safe environment for future generations, farmers should adopt alternative methods of pest control, such as Integrated Pest Management (IPM) and organic farming.

---

* Corresponding Author's Email: sushilkumar832012@gmail.com
† Corresponding Author's Email: saniya.tyagi13@bhu.ac.in

In: Insecticides
Editors: Archana Anokhe and Govindaraj Kamalam Dinesh
ISBN: 979-8-90134-022-6

Additionally, opt for organic food products to avoid the harmful effects of pesticides.

Keywords: agriculture, disorder, environmental pollutants, insecticide, human health

Introduction

Suppressing insect pests is a key component of sustainable agriculture, which is primarily based on the use of insecticides. According to the Environmental Protection Agency (EPA), pesticides are a class of chemicals that control and repel insect populations. These are chemical and natural agents used to control or eliminate pests, including animals, organisms that cause plant diseases, and vegetation. It also controls other organisms, such as nematodes, arthropods (excluding insects), and vertebrates, which damage our food sources and cause numerous health issues (Rani et al., 2021). In addition, these pesticides save man's energy and time and are highly effective (Igbedioh, 1991; Boxall, 2001; Narayanasamy, 2006; Cooper and Dobson, 2007; Schreinemachers et al., 2017; Wang et al., 2017).

Currently, insecticides are considered an indispensable tool by farmers to ensure the maintenance of crop production, both in terms of quality and quantity, to meet the demands of an ever-increasing human population.

Before 1940, certain chemicals were available, specifically botanical insecticides such as pyrethrins, nicotine, and rotenone (derived from derris); however, their use was limited due to their rapid deterioration in sunlight. To mitigate these challenges, humans resorted to the application of synthetic pesticides. Sought inexpensive chemicals that could persist in sunlight, have low toxicity to humans, and swiftly exterminate insect pests. The discovery of Dichlorodiphenyltrichloroethane (DDT), the premier synthetic organochlorine insecticide, was particularly sought after for its comprehensive efficacy against insect pests that had a significant impact on agriculture and public health. Usage in India started in 1948, with the import of DDT for malaria control and BHC for locust control. DDT commenced the discovery of other synthetic pesticides and their broad adoption in the country.

Only later did problems associated with insecticidal persistence in the environment be recognized and brought to light in the public domain (Carson, 1962). The use of synthetic pesticides at higher trophic levels, including humans, has resulted in various health problems and diseases due to chronic exposure.

Environmentalists and consumers have been concerned about pesticide residues in food due to the widespread use of pesticides. The majority of pesticides, particularly organochlorines, are highly resistant to microbial degradation (Ejobi et al., 1996). can accumulate in the human body and the environment, posing a threat to human health. Due to their persistence and lipophilicity, pesticides and their residues can be concentrated in the adipose tissues and blood serum of animals, resulting in environmental persistence, bioconcentration, and biomagnification along the food chain (Darko and Acquaah, 2007). Insecticides have been widely used in the past and have been associated with a variety of adverse health effects in humans, including endocrine disorders (Lemaire et al., 2004), effects on embryonic development (Tiemann, 2008), lipid metabolism (Karami-Mohajeri and Abdollahi, 2011), and haematological and hepatic alterations (Freire et al., 2015). Now, since the use of organochlorides has been restricted in many parts of the world, other classes of insecticides, such as organophosphates, synthetic pyrethroids, neonicotinoids, and fumigants like aluminium phosphide, pose a significant threat to farmers. Their exposure causes poisoning and other health risks, even leading to death in severe cases.

Pesticides enter the body through the skin, mouth, eyes, and respiratory system. As a result, several acute diseases have been scientifically proven to be linked to pesticides, including headaches, nausea, vomiting, skin rashes, respiratory disorders, eye irritation, sneezing, convulsions, and coma (Raschke and Burger, 1997). can potentially be fatal when used directly (Page et al., 2017; Dasgupta et al., 2005).

Pesticide toxicity is mainly influenced by oral exposure. Chronic diseases and disorders are often associated with the regular consumption of food crops grown in pesticide-treated soil (Macharia, 2015). Acute pesticide poisoning is now a very uncommon occurrence, and long-term subclinical effects are still a problem. Exposure to low concentrations of pesticides can lead to chronic poisoning, which may not be apparent for years (Capoferri, 2018). Diseases include cancer, asthma, dermatitis, endocrine problems, infertility, immunotoxicity, neurobehavioral disorders, and congenital abnormalities (Kim et al., 2017). The primary effect on cells and DNA damage might affect cellular homeostasis, leading to chronic diseases (Alleva et al., 2018). Moreover, pyrethroid concentrations in the environment cause carcinogenic processes by influencing genetic and molecular alterations (Caron-Beaudoin et al., 2018; Navarrete-Meneses et al., 2019).

People exposed to pesticides at work, such as farm and rural workers, experienced both acute and chronic impacts (Gangemi et al., 2016). Insecticide-induced poisoning, its potential risks in humans, and its management using antidotes have become crucial. In this chapter, we will discuss insecticides and their routes of entry into the human body, their adverse health effects in both acute and chronic poisoning, and their treatment using antidotes, including their mode of action.

Routes of Exposure to Insecticides

The term "pesticide exposure" refers to the situation where a surface or organism comes into contact with pesticides. A person's body can be exposed to pesticides through contact or ingestion. The duration of exposure and the concentration of the pesticides present are important factors in determining the overall toxicity of the effect. Can be exposed to pesticides in one of four ways: either inhalation (when they breathe in a pesticide), dermal (through their skin), ocular (through their eyes), or oral (when they swallow a pesticide) (Kumar et al., 2012).

Inhalation

Respiratory exposure, which involves inhalation or breathing, is a crucial concern in the context of pesticide usage, and the volatile components of these chemicals are the primary cause of this exposure (Msibi et al., 2021; Chen et al., 2022). Ingestion of insecticides in significant quantities can lead to severe damage to the nose, throat, and lung tissues (Damalas and Eleftherohorinos, 2011). The probability of insecticide exposure is generally low when conventional application equipment is used to spray large droplets. However, using low-volume equipment to apply a concentrated substance leads to an increased potential for respiratory exposure due to the production of smaller droplets (Amaral 2014).

It is essential to recognize that inhaling large quantities of pesticides in the environment through air, water, or soil poses a significant health risk to workers, particularly affecting their nose, throat, and lung tissues (Chen et al., 2022).

The evaluation of household exposure to pyrethroids through all exposure pathways was discussed in Japanese children (Yoshida et al., 2021). A collection of urine excreted by subjects aged 6 to 15 years (n = 132) upon waking up, and the sampling of airborne pyrethroids in the subjects' bedrooms for 24 hours was carried out. Contribution rates of the quantities absorbed via inhalation relative to the quantities absorbed via all exposure pathways tended to decrease in the following order: profluthrin (15%), transfluthrin (14%), metofluthrin (1%), and bifenthrin (0.1%). It was the most prominent pyrethroid as a household air pollutant, as exposure to transfluthrin resulted in adverse health effects.

Dermal Contact

Dermal exposure is widely acknowledged as the predominant and efficacious mode of exposure (Krieger et al., 2000) for farmers who are exposed to insecticides through spillage, splashing, or spray drift (Bootsikeaw et al., 2021), particularly when they use insecticides in agriculture or household activities (Siriwat et al., 2021).

Absorption of insecticides may also result from exposure to significantly large amounts of residue. Formulations vary widely in their physical and chemical properties and ability to be absorbed through the skin (Beard et al., 2014). can be affected by the amount and length of exposure, other materials on the skin, temperature, humidity, and the use of personal protective equipment (Macfarlane et al., 2013).

Wang et al., (2021) used absorbent paper patches to directly examine dermally exposed workers to Trinexapac-ethyl (TE). Exposure intensity (EI) and potential dermal exposure (PDE) for each body part of mixing/loading, manual sprayer, and handheld power sprayer on people ranged from 7.22 to 73.0 pg/cm^2. The upper arms of mixing and loading workers had the lowest EI of TE, while manual sprayers had the highest. Loading workers' chests and backs were 29% exposed, handheld power sprayers 40%, and thighs through manual sprayers 32%. et al., (2021) investigated the possibility of dermal exposure to pesticides and evaluated the risk of such exposure for those involved in the seed coating process. The study found that the degree of chemical exposure varied among workers, depending on the type of work they performed and the part of their body exposed to the chemical. Handling of tebuconazole during the seed-coating process posed little danger to human health. In contrast, the handling of carbofuran posed a potential threat.

Ocular

The ocular tissues are at a considerable risk of chemical injury. Pesticides have been determined to be absorbed by the eyes in adequate quantities, leading to severe or fatal illnesses (Gilden et al., 2010). The size and weight of individual particles contribute to the specific hazard posed to the eyes by granular pesticides (Jaga and Dharmani, 2006). Power equipment is utilized to apply pesticides; however, pellets may bounce off vegetation or other surfaces with great force, resulting in significant eye damage (Fareed et al., 2012). Protection is also necessary when handling concentrated or highly toxic pesticides during measuring or mixing. To prevent eye contact with dust, protective face shields or goggles must be worn while spraying pesticides.

Ingestion

The occurrence of oral exposure to a pesticide often arises by accident or intentionally (Damalas and Eleftherohorinos, 2011). Most accidental cases of oral exposure arise when pesticides are transferred from their original, labeled containers to an unlabeled bottle or food container (Gilden et al., 2010). to pesticides through the oral route may occur when individuals engaged in the production or utilization of pesticides fail to wash their hands prior to consuming food or smoking (Kwong et al., 2021; Wang et al., 2021). The general public may be exposed to oral exposure when consuming food contaminated with pesticides (Yohannes et al., 2014). Incidents have been reported where individuals have suffered from poisoning due to the consumption of pesticides stored in soft drink bottles or water that was contaminated by pesticides (U.S. Environmental Protection Agency, USEPA, 2007). Exposure is the most common route of poisoning that causes several health problems in humans (Lei et al., 2015; Shi et al., 2013).

Effect of Insecticides on Human Development and Reproduction

The synthesis and use of insecticides were primarily intended to serve humankind. By nature, they have long helped humans battle their food crisis by controlling insects in agricultural fields, in public health by controlling many vector-borne diseases such as malaria, dengue, and cholera, and by controlling household pests like cockroaches. Despite their successful applications in many sectors, they appear to have a negative impact on the environment and non-target organisms. Then, cases of pest resistance and insecticide resistance are observed in agriculture. The effect of these chemicals on human health is enormous. Insecticides can enter the human body by one of four means:

1. Dermal exposure during mixing of insecticides, spray drift, spills, and other factors, such as improper handling
2. Oral ingestion can occur accidentally during the inhalation of spray vapors, when containers of insecticides are used to store food materials, or when hands are not washed after handling such chemicals.
3. Respiratory exposure via low-volume sprays is now used in fields by farmers.
4. Eye exposure during sprays or by drift (Kumar and Kumar, 2019).

Insecticides, after reaching the body, cause discomforts like stinging eyes, rashes, blisters, skin irritations, blindness, nausea, dizziness, diarrhea, etc. (Shah, 2020), along with many diseases in humans like- leukemia, malignancies and asthma (Kumar and Kumar, 2019), in severe cases, it may also result in the death of the person (Shah, 2020).

In the 1970s, pesticide poisoning was responsible for the deaths of about 50,000 people, and several million were otherwise affected (Fedorov and Yablokov, 2004). The first case of insecticide resistance reported in India was that of Parathion in Kerala in 1958, where approximately 100 individuals died due to contaminated wheat flour (Kumar and Kumar, 2019). This is a perfect example that shows people who are not directly linked to pesticide production, packaging, or any other work can be poisoned due to the faulty delivery of pesticide-loaded products (Fedorov and Yablokov, 2004).

It is essential to recognize that a dose-effect relationship has been established for the majority of pesticides and that the use of biochemical measurements can facilitate the identification of pesticide effects before the onset of clinical health detriments. There may exist a threshold level below which no effects are observable, commonly referred to as the "no-observed-effect-level." Suppose pesticides are suspected of causing carcinogenic activity. In that case, it may not be relevant to detect preliminary minor effects and establish a no-observed-effect level.

The poisoning due to organophosphates in school children of rural areas where agriculture or related activities are conducted results in low IQ of children and lower processing speed as compared to other children of their age. Diseases caused by changes in striatal dopamine activity (Fedorov and Yablokov, 2004) are also linked to pesticide exposure. Immunity due to occupational status is more commonly observed in men than in women (WHO, 1990). Moreover, kidney disorders are also seen, as these organs are primarily responsible for detoxifying the body. The body is subjected to toxins, which severely affect it (Kumar and Kumar, 2019). Brain tumors in children are more common when their parents, especially the mother during pregnancy, are exposed to indoor insecticides or garden insecticides, which may also be due to residential effects. Cancer in females and prostate cancer in males are also caused by insecticides disrupting the endocrine system and hormones.

The reproductive effects are also seen due to insecticides, such as less testosterone production leading to less sperm production, a short gestation period, neurological disorders in the fetus, retardation in fetal growth, and pregnancy loss (Shah, 2020). Cases related to stillbirths are not unheard of in instances of pesticide toxicity in mothers. The insecticides are also found to overcome the placental barrier, are transferred from the mother to the baby through breast milk, and are stored in the subcutaneous tissues of the infant (Fedorov and Yablokov, 2004). Other than occupational and other indirect exposures to insecticides, there is yet another reason for the death of a person: the intentional use of pesticides for suicide attempts due to their easy availability (WHO, 1990).

Effect of Acute Poisoning on Human Health

Acute toxicity can be defined as a chemical's or pesticide's capacity to cause systemic damage as a consequence of a single exposure to a relatively high amount of the chemical (Nesheim et al., 2012). exposure can cause acute illnesses or injuries within 24 hours.

Acute pesticide poisoning symptoms include numbness, lack of coordination, tingling sensations, headache, nausea, tremors, disorientation, abdominal cramps, impaired vision, sweating, difficulty breathing, or slow heartbeat (Kumar et al., 2012).

Acute pesticide-related disorders are prevalent among agricultural workers in food-producing countries due to the use of toxic substances such as insecticides and nematicides. Organophosphates, carbamates, organochlorines, pyrethrum, and pyrethrin are among the major categories used. There are various pesticide formulations containing numerous active ingredients. Acute organophosphate and carbamate poisoning includes salivation, incontinence, convulsions, muscle twitching, diarrhea, visual disturbances, respiratory difficulties, and rapid heart rate. An excess of the neurotransmitter acetylcholine is responsible for these symptoms, as this substance attacks a chemical called cholinesterase (Mergler, 1998).

Table 1 presents the measures and warnings related to acute toxicity. Acute toxicity tests are frequently utilized to ascertain the impact of high chemical exposure. The outcome reveals a high degree of fatality (USEPA, 1994). Long-term chronic toxicity tests are conducted to determine the effects of exposure to relatively low-toxic concentrations. The result specifies the sub-lethal effect, such as reproduction and growth, or both lethality and sub-lethal effects (USEPA, 1994).

Pesticides possess the potential to cause harm to both humans and animals through acute toxicity resulting from a single exposure of brief duration. Exposure can occur through various routes, including dermal, inhalation, oral, and ocular. To determine acute toxicity, test animals were assessed for dermal, inhalation, and oral toxicity. Evaluations of eye and skin irritations are taken into account during the examination of the test subjects (Nesheim et al., 2012; Hock and Lorenz, 2006).

Table 1. Acute toxicity measures and warnings

Categories	Signal word	Categories of acute toxicity			Oral lethal dose
		LD_{50}	LD_{50}	LD_{50}	
		Oral (mg/kg)	Dermal (mg/kg)	Inhale (mg/L)	
Highly toxic	Danger poison (skull and crossbones)	0-50	0-200	0-0.2	A few drops to a teaspoonful
Moderately toxic	Warning	50-500	200-2,000	0.2-2.0	Over a teaspoonful to one ounce
Slightly toxic	Caution	500-5,000	2,000-20,000	2.0-20	Over one ounce to one pint
Relatively non-toxic	Caution (or no signal word)	5,000+	20,0000+	20+	Over one pint to one pound

Source: (Nesheim et al., 2012).

The term LD50 is commonly used to refer to acute toxicity, and it means 50% of the test animals were adversely affected by the dose under controlled conditions (Fischel et al., 2011). is quantified in milligrams (mg) of pesticide per kilogram (kg) of body weight and has been widely accepted as a benchmark for determining acute toxicity (USEPA, 2012). The dose, whether administered through ingestion or injection, is given to test subjects, who are then closely monitored for a specified period.

The LD50 has made it feasible to compare the relative toxicities of pesticides, as the values obtained are standardized measurements according to the norms (USEPA, 2012). The toxicity of a pesticide is directly proportional to its LD50 value, such that lower values indicate higher toxicity and vice versa (USEPA, 2012). Analysis of data can provide insights into the concentration at which 50% of organisms exhibit a particular effect, commonly known as the median effective concentration (EC50). In cases where acute toxicity tests report EC50, the test results also confirm the species involved, the duration of the test, and the life cycle stage of the test species. Its counterpart, LC50, EC50, is a point estimate, and as such, a coefficient of variation can be determined (USEPA, 1994).

Effect of Chronic Poisoning on Human Health

Chronic diseases, characterized by their generally slow progression and long duration, are the primary cause of mortality in the New World, accounting for over 60 % of deaths. According to the WHO, out of the 36 million people who passed away in 2008 from chronic disease, nine million were under the age of 60, and 90% of these premature deaths occurred in low- and middle-income countries (Mostafalou and Abdollahi, 2013).

The chronic health effects described by Ostrea et al., (2009) exhibit a latency period, which refers to the interval between initial exposure and the onset of the disorder. Effects encompass a range of conditions, including leukemia, cancer, lung diseases, asthma, cirrhosis, kidney and urinary tract diseases, obesity and diabetes, gastrointestinal disorders, cardiovascular and hematologic diseases, genetic alterations, as well as pathological states affecting the bone marrow blood cells and central nervous system (Garg et al., 2004; Carozza et al., 2009; Franco et al., 2010).

Awareness of neurotoxic chemical hazards has caused countries to reduce exposure levels. The long-term exposure levels for most chemicals remain unknown. Exposure to low to medium levels of neurotoxic substances over time can slowly alter nervous system functions. Alterations may go unnoticed in the early stages. The nervous system can compensate in the beginning stages through new learning (Mergler, 1998).

Antidotes of Pesticide Poisoning and Their Mechanism of Action

Antidotes counteract the toxicity/ poisonous effects of pesticides in humans by various mechanisms. A suitable antidote for a particular insecticide is crucial for its timely administration to nullify its lethal effects. In a developing country like India, most lethal poisoning cases are deliberate through organophosphates. (OP) Induced nerve poisoning has long been treated with atropine and oximes, which counteract muscarinic effects (Cannard, 2006; Thiermann, 2013).

Atropine improves cardiorespiratory function and quickly suppresses muscarinic effects. It is administered intravenously (adults receive 1-3 mg, children receive 0.02 mg/kg, and is continued in doubling doses every five minutes until the bronchospasm and bronchorrhea have stopped and cardiovascular function has been recovered (Abedin et al., 2012). can be administered continuously to maintain a sufficient level of cardiorespiratory function. A massive dose of atropine may be necessary for severe poisoning. The infusion should be temporarily discontinued and reintroduced at around 70% of the prior rate until toxicity has subsided if there is an indication of atropine poisoning (dry mouth, tachycardia, agitation or disorientation, dry skin, ileus) (Eddleston, 2020). Patients must be closely monitored for the return of cholinergic symptoms necessitating bolus atropine therapy.

If oximes are not administered after enzyme aging, they can revive phosphorylated cholinesterases. Their efficacy has not been shown; they can increase the patient's consciousness, decrease fasciculations, and restore muscle power (Blumenberg et al., 2017). Chloride (or obidoxime) should be administered, preferably in a critical care setting, as a loading dosage of 30 mg/kg (250 mg obidoxime) over 30–60 minutes, followed by an infusion of 8–10 mg/kg/hour (32 mg/hour obidoxime) for 2-3 days. Daily withdrawal attempts can be reintroduced if muscle weakness worsens after discontinuing oximes.

Conventional treatment does not work well in many cases of OP pesticide and nerve agent poisoning (Jokanovic and Stojiljkovic, 2006; Cherain et al., 2005). This has led to extensive research to find new antidotes that work better or can be used in conjunction with conventional treatment. The development of a combined therapy using the muscarinic receptor antagonist atropine and an oxime for the reactivation of organophosphate (OP)- inhibited acetylcholinesterase (AChE) was prompted by the need for an efficient remedy for OP poisoning. In addition to the commonly used oxime pralidoxime (2-PAM), other more efficient oximes include charged and non-charged compounds, mono- and bispyridinium oximes, asymmetric oximes, oximes with different substitutes, and, more recently, non-oxime reactivators of OP-inhibited AChE and to counter OP-induced cholinergic signs (Worek et al., 2016).

Other types of pesticides that frequently result in serious and/or fatal poisoning include carbamate and organochlorine insecticides, fumigants such as aluminum phosphide (a significant issue in northern India), and herbicides like paraquat. Herbicides, chlorphenoxy acetic acid derivatives, propanil, several pyrethroid insecticides, avermectins, and amitraz are less frequent causes of severe poisoning (Sharma et al., 2015). The effects of poisoning from carbamate insecticides are comparable to those of organophosphorus pesticides. Carbamate-inhibited acetylcholinesterase reactivates rather quickly; carbamate poisoning is typically less severe and lasts for a shorter period. There are fewer reports of chronic consequences. Fluids and atropine therapy should be combined with supportive measures and resuscitation as necessary. They are rarely necessary. Management is primarily supportive and symptomatic because there are no antidotes for newer classes of pesticides, like pyrethroids and neonicotinoids. A patient presenting with acute poisoning requires their airway, breathing, and circulation to be optimized (Ramchandra et al., 2019). atropine may be thought to reduce salivation, but it is likely to have negative consequences since it may make patients with pyrethroid poisoning more susceptible to seizures (Drago et al., 2014).

Factors Affecting the Toxicity of Insecticides

Pesticides, including insecticides, can affect humans directly and indirectly via various routes. Penetrate the human body through the skin, mouth, eyes, and respiratory system (Kim et al., 2017). Exposure to pesticides is the key factor determining their toxicity (Quijano et al., 2016), whether through direct ingestion of pesticides or via contaminated food. It is the primary source of humans' direct consumption of toxic substances (Kalyabina et al., 2021). becomes important when the waiting period for the harvest is ignored. Consumption of food based on crops grown on soil contaminated with pesticides results in short-duration (acute) and long-duration (chronic) diseases and disorders by accumulating pesticides in their edible and inedible parts in concentrations that are high enough to cause clinical problems in animals and humans (Macharia, 2015; Yousaf et al., 2016). It is thus necessary to wash the food items before consuming them.

The toxic effect of pesticide exposure depends on the amount of pesticide involved and the time of its persistence (Singh et al., 2018). In our daily routine, we are exposed to pesticides several times, either directly or indirectly (USEPA 2016). Exposures occur either through dietary products, such as food and drinking water, or the application of insect repellents in our homes or on our skin. Exposures occur through occupational exposure, such as farmers working in open fields and greenhouses, workers in the pesticide industry, and exterminators of house pests. Regardless of the occupation, the presence of such pesticides in the workplace also contributes to potential occupational exposure. Those who mix, load, transport, and apply formulated pesticides are considered to receive the most significant exposure due to their work and, therefore, are at the highest risk for possible acute intoxications (Valenzuela et al., 2009).

In some cases, exposure may occur due to accidental spills of chemicals, leaks, or faulty spraying equipment. Exposure also increases if farmers do not follow instructions carefully about applying pesticides. When they ignore basic safety guidelines, such as using personal protective equipment and fundamental sanitation practices like washing hands after handling and applying pesticides or before eating (Fenske and Day, 2005), factors can affect pesticide exposure during handling (Fenske and Day, 2005). Forms of pesticide formulation may affect the degree of exposure. Formulations are easily susceptible to direct contact through splashing and occasional spillage, as well as indirect skin contact through clothing contamination. Formulations like dust can result in exposure to the face and eyes, and also pose respiratory hazards when being loaded into the application equipment. On the type of packaging, some extent of pesticide exposure also occurs during the opening of the bag. The size of packaging containers, such as cans and bottles, may pose a risk of spillage and splashing. Adjuvant chemicals used in the preparation of formulations to enhance their activity and facilitate application to reach target species may show toxicity themselves, thus contributing to the overall effect of exposure to a commercial pesticide product (Surgan et al., 2010).

Toxicity depends on the chemical nature of the insecticides, primarily due to their lipophilicity and hydrophobicity, which induce chronic effects. They are typically distributed throughout the organism, including humans, by binding to plasma proteins, blood cells, and lipids in various organs and peripheral tissues (Clark, 2018). The lipophilicity of molecules determines the strength of binding. The data on the water solubility of toxicants provide the basis for estimating their bioaccumulation, biodegradation, hydrolysis, and adsorption (Winkler 2005). The lipophilicity of pesticides increases both their effectiveness in pest control and their subsequent bioaccumulation, as seen for organochlorides, especially DDT (Ali et al., 2016). The hydrophobicity and chemical structure of both the starting product and pesticide play an important role (Iwafune, 2018).

Not practicing safe application techniques for pesticides by applicators is also a significant cause of their accidental exposure. Incorrect direction of spraying/dusting, mishandling of chemicals, failure to wash hands before eating and drinking, and the use of old or inadequate pesticide application equipment may pose accidental exposure or serious health issues in humans. It becomes crucial to gain a thorough understanding of pesticide handling and its proper application to reduce such mishaps.

Safety Precautions Are Followed During Mixing, Spraying, and After Use

Storage of Pesticides

Pesticides must be stored safely due to their potential risks. Moreover, chemicals should be stored separately. Pesticide stores are inspected annually to ensure safety. Locations should avoid flooding and have measures in place to prevent spills. The building must be secure, dry, and frost-free, with appropriate warning signs and staff training (Graham 2015).

Guidelines for Pesticides Stored on Farms

Avoid excessive stock by maintaining only the amount needed for the near future. It should be kept in a locked and well-ventilated area. It should be prevented. It should be out of reach of children. It should not be stored near food, animal feed, or medical supplies. Should be kept in their original containers along with the label and instructions. It is essential to follow the storage instructions provided on the label. Transferring pesticides to other containers may lead to confusion among children or others who may mistake them for food or drink. When using products at home, ensure that child-resistant containers are used and tightly sealed after each use (Graham 2015).

Precautions for Operators

If the operator feels giddiness or uneasiness, they must discontinue spraying/dusting at once. Should not spray/dust more than 4 hours at a stretch in a day and should not take up spray/dusting work on an empty stomach (Prasad 2016).

Precautions to be Followed Before Applying the Pesticide

To effectively manage pests, it is essential to have a comprehensive understanding of the specific pest and the extent of damage it causes. Utilize pesticides when necessary, and seek expert guidance on the most appropriate methods of control. It is imperative to use the recommended pesticide for the given problem. If multiple options are available, opt for the least toxic to mammals and, if feasible, the least persistent. Read the label, including the finer details, and ensure that the proper protective clothing is accessible and worn and that the individual applying the pesticide is fully trained in proper application techniques. All application equipment for leaks should be calibrated with water and confirmed to be functioning optimally. Ensure they have sufficient water, soap, towels, and clean clothing on hand. It is vital to avoid inhaling pesticides, and they should always be stored in a locked area. Neighbours of any spray program, particularly those with apiaries. Only the necessary amount of pesticide for daily application, and store it on-site without transferring it to other containers (Prasad T. V. 2016).

While Mixing or Applying a Pesticide

Avoid smoking and eating while mixing or applying pesticides, as you can easily transfer pesticides from your hands to your mouth. The label's "direction for use" states that it is to be used only for the specified purpose. The amount will not do twice the job. May harm humans, others, the plant, location, or pet you are protecting. An apron and face shield can prevent splashes on the coveralls or face. The risk of inhaling small droplets during mixing and loading is low, and minimizing inhalation exposure is recommended (Wolf et al., 1999).

Always mix or dilute pesticides in a well-ventilated area outside or in a well-ventilated environment. Only use the indicated amount and measure insecticides carefully. Reuse measuring cups or spoons from the kitchen and mix only the amount needed for each application. Do not prepare a larger amount of spray solution for storage in case of future use. Pesticides will degrade and become ineffective over time. Children, pets (including birds and fish), and toys (including pet toys) should be kept away from areas where you mix and apply pesticides for at least the length of time required on the label (Prasad T. V. 2016).

Suppose the label does not indicate a specific duration. In that case, it is advisable to wait until the pesticide has completely dried before reentering the treated area. Suppose clothing becomes drenched in the spray solution or contaminated with the pesticide concentrate during the spraying process. In that case, it is recommended that it be disposed of immediately (Hall et al., 2002). Areas of the body that may have come into contact with the pesticide should be thoroughly washed with soap and water (Landers, 2004). It is also recommended that fresh, clean clothing be worn after exposure to the contaminated area. When placing insect or rodent baits, it is crucial to ensure that they are not accessible to small children or pets, whether indoors or outdoors. To prevent harm to non-target organisms, it is advisable to use selective insecticides that have a restricted range of target pests (Prasad T.V. 2016).

Precautions After Application

Proper post-spraying care is also essential. Any pesticide product that remains unused must be returned to the store.. It is essential to wash all equipment after use. To do so thoroughly may result in the accumulation of residues, which can be dangerous if they reach a watercourse (Ramwell et al., 2004). Precise calibration allows the diluted spray to be utilized during field treatment, enabling tank washings to be done in the treated area. The sprayer tank washings can be applied by spraying on the final swath, making extra water accessible in the field. Areas called 'biobeds' are used in some countries for disposing of diluted pesticide-contaminated water. It can be a lined pit filled with a mixture of straw, peat, and soil and covered with turf grass (Fogg and Carter 1998; Henriksen et al., 2003; Basford et al., 2004).

Fogg et al., (2003a) found that biobeds can degrade some herbicide and insecticide mixtures, but certain pesticides should be avoided. Water management is crucial for lined beds to prevent microbial degradation and ensure proper drainage without overflows. Beds remove over 99% of mobile pesticides within 9 months, but leaching can occur (Fogg et al., 2003b). Biobeds contain only 0.1% of the pesticides found in run-off from concrete surfaces. Farm-scale effluent treatment plants can remove small amounts of pesticides from water (Harris et al., 1991).

Emergency Measures

Signs and Symptoms of Poisoning

Poisonings due to pesticides are usually acute and result from extensive skin contact or ingestion. Symptoms vary with the type of pesticide and can sometimes be confused with those of other illnesses.

Indications of Pesticide Poisoning

- General: extreme weakness and fatigue.
- Skin irritation, burning sensation, excessive sweating, and staining
- Eyes: itching, burning sensation, watering, complex or blurred vision, narrowed or widened pupils.
- Digestive system: burning sensation in the mouth and throat, excessive salivation, nausea, vomiting, abdominal pain, diarrhea.
- Nervous system: headaches, dizziness, confusion, restlessness, muscle twitching, staggering gait, slurred speech, fits, unconsciousness.
- Respiratory system: cough, chest pain and tightness, difficulty with breathing, wheezing.

It is essential to gather additional information. An effort should be made to obtain evidence from pesticide containers or spray equipment, and the labels on these containers should be read and retained. Pesticide poisoning is suspected; first aid must be given immediately, and medical advice and help must be sought at the earliest opportunity. If possible, the patient should be taken to the nearest medical facility.

First-Aid Treatment

When a spray operator becomes ill while working, the doctor should be informed of the active ingredient and given information about the chemical. Depending on the type of poisoning, atropine injection helps with OP or carbamate poisoning, but there are no organochlorine antidotes. Fuller's earth can help a person who has ingested paraquat. It is not appropriate for patients poisoned by pesticides (Graham 2015).

If breathing has stopped, give artificial respiration. No insecticide has been swallowed. Mouth-to-mouth resuscitation may be given. The patient's chin is up, and the head is tilted back with one hand to keep the airway clear. The other hand is on the patient's forehead, with the thumb and index finger toward the nose. Together, the patient's nostrils with the thumb and index finger to prevent air from entering the lungs. Adequate respiration is taking place, and the chest should rise and fall. Your mouth allows the patient to breathe out. Take another deep breath, form a tight seal around the patient's mouth, and blow into it again. This procedure is 10-12 times a minute (once every five seconds). Respiration should be continued for as long as possible if a pulse is still present. An insecticide has been swallowed; another form of artificial ventilation should be used. Administer a deep breath, then form a tight seal with your mouth over and around the patient's mouth. Breathe four quick, full breaths in first without allowing the lungs to fully deflate, and watch the patient's chest (Prasad, 2016).

In many countries, specialized centers exist that cater to poison-related concerns, which medical practitioners can utilize for treatment advice. Several tropical countries lack the necessary proficiency in this domain (Ngowi et al., 2001a), despite being locations where some of the most lethal pesticides are used, often with substandard sprayers that are prone to leakage and operated by untrained personnel. Occupational poisoning has persisted as a severe problem among farm labourers, particularly those working in coffee plantations in Tanzania (Ngowi et al., 2001b).

Vomiting

Do not induce vomiting unless the patient has swallowed a pesticide that is known to be highly toxic, and medical help is not expected soon. Induce vomiting if the patient has swallowed oil spray or products diluted in diesel or kerosene because of the possibility of inhalation of the vomited material, which would be more dangerous than intestinal poisoning. The product label should indicate whether the pesticide is highly toxic (skull-and-crossbones signs). It should be induced only if the patient is conscious. Necessary to sit or stand the person up and tickle the back of the throat with a finger. Whether vomiting occurs or not, give the patient a drink consisting of three tablespoons of activated charcoal in half a glass of water until medical help arrives (Prasad T. V., 2016).

If There Is Insecticide on the Skin or in the Eyes

Rinse the eyes with large quantities of clean water for at least five minutes. Contaminated clothing from the patient should be removed, and the patient should be removed from the contaminated area. The body should be completely submerged for at least 10 minutes, using soap if possible. No water is available; gently wipe the skin with a cloth or paper to absorb the pesticide, and avoid harsh rubbing or scrubbing.

Caring for the Patient

Make the patient lie down and rest because poisoning with organophosphorus and carbamate compounds is made worse by movement. The patient is lying on their side with their head lower than their body. If the patient is unconscious, pull the chin forward and the head back to ensure a clear airway. The patient should be wrapped in a blanket if they feel cold, and the patient's temperature should be cooled by sponging with cold water if excessive sweating occurs. The patient vomits spontaneously; ensure that they do not inhale the vomit. In the event of convulsions, put padded material between the teeth to avoid injury. Do not allow patients to smoke or drink alcohol. Do not give milk. It can be given.

Conclusion

It is important to note that the severity of these health impacts can vary depending on the type of insecticide used, the level and duration of exposure, and the age and health of the person exposed. It is always recommended to follow the instructions on insecticide labels carefully and to take proper safety precautions when handling or using insecticides. There should be a focus on developing newer, more efficient antidotes, especially for synthetic pyrethroids and neonicotinoids.

References

Abedin MJ, Sayeed AA, Basher A, Maude RJ, Hoque G, Faiz MA. Open-label randomized clinical trial of atropine bolus injection versus incremental boluses plus infusion for organophosphate poisoning in Bangladesh. *of Medical Toxicology* (2012) 8: 108e17.

Ali N, Rajeh N, Wang W, Abualnaja KO, Kumosani TA, Albar HMS, Eqani SAMAS, Ismail IMI. Organohalogenated contaminants in type 2 diabetic serum from Jeddah, Saudi Arabia. *Pollution* (2016) 213:206–212, https://doi. org/10.1016/j.envpol.2016. 01.087.

Alleva R, Manzella N, Gaetani S, Bacchetti T, Bracci M, Ciarapica V, Monaco F, Borghi B, Amati M, Ferretti G, Tomasetti M. Mechanism underlying the effect of long-term exposure to low dose of pesticides on DNA integrity. *toxicology* (2018) 33:476–487, https://doi.org/10.1002/tox.22534.

Amaral AFS. and Asthma: Challenges for Epidemiology Front Public Health. 2014) 2: p.6.

Basford WD, Rose SC, Carter AD. On-farm bioremediation (biobed) systems to limit point source pesticide pollution from sprayer mixing and wash-down areas. *of Applied Biology* (2004) 71:27–34.

Beard JD, Umbach DM, Hoppin JA, et al., Pesticide exposure and depression among male private pesticide applicators in the agricultural health study. *Health Perspectives* (2014) 122(9):984–991.

Blumenberg A, Benabbas R, deSouza IS, et al., Utility of 2-pyridine aldoxime methyl chloride (2-PAM) for acute organophosphate poisoning: a systematic review and meta-analysis. *of Medical Toxicology* (2017) 14: 91e8.

Bootsikeaw S, Kongtip P, Nankongnab N, Chantanakul S, Sujirarat D, Mahaboonpeeti R, Khangkhun P, Woskie S. Urinary Glyphosate Biomonitoring of Sprayers in Vegetable Farm in Thailand. *and Ecological Risk Assessment* (2021) 27: 1019–1036.

Boxall RA. losses to insects—a world overview. *Biodeterioration and Biodegradation* (2001) 48(1-4): 137-152.

Cannard K. acute treatment of nerve agent exposure. *of the Neurological Sciences* (2006) 249: 86e94.

Capoferri D, Della Pelle F, Del Carlo M, Compagnone D. Affinity sensing strategies for the detection of pesticides in food.

Caron-Beaudoin E, Viau R, Sanderson JT. Effects of neonicotinoid pesticides on promoter-specific aromatase (CYP19) expression in Hs578t breast cancer cells and the role of the VEGF pathway, *Environmental Health Perspectives* (2018) p.126.https:// doi.org/10.1289/EHP2698.

Carozza SE, Li B, Wang Q, Horel S, Cooper S. Agricultural pesticides and risk of childhood cancers. *Journal of Hygiene and Environmental Health* (2009) 212: 186-195.

Carson, R. "Silent Spring Hamish Hamilton." (1962).

Chen DW, Li JG, Zhao YF, Wu YN. Exposure of Fipronil Insecticide and The Associated Health Risk. *of Agricultural and Food Chemistry* (2022) 70: 63–71.

Cherian MA, Roshini C, Peter JV, Cherian AM. in organophosphorus poisoning. *Journal of Critical Care Medicine* (2005) 9:155-163.

Clark RD. mammalian metabolism and toxicity of pesticides in silico. *Management Science* (2018) 74:1992–2003, https://doi.org/10.1002/ps.4935.

Cooper J, Dobson H. benefits of pesticides to mankind and the environment. *Protection* (2007) 26(9): 1337-1348.

Crauford-Benson HJ. Naples typhus epidemic 1943–4. *British Medical Journal* (1946) 1: 579–680.

Damalas CA, Eleftherohorinos GE. exposure, safety issues, and risk assessment indicators. *Journal of Environmental Research and Public Health* (2011) 8(5):1402–1419.

Darko G, Acquaah SO. of organiclorine pesticides residues in meat. *Journal of Environmental Science and Technology* (2007) 4: 521-524.

Dasgupta S, Meisner C, Huq M. Health Effects and Pesticide Perception as Determinants of Pesticide Use: Evidence from Bangladesh. *World Bank.* 2005).

Drago B, Shah NS, Shah SH. Acute permethrin neurotoxicity: variable presentations, high index of suspicion. *Reports* (2014) 1:1026–1028.: 10.1016/j.toxrep.2014.09.007.

Eddleston M. Poisoning by pesticides. *Medicine* (2020) *48*(3): 214-217.

Ejobi F, Kanja LW, Kyule MN, Muller P, Kruger J, Latigo AAR. Organoclorine pesticides residues in mothers milk in Uganda. *of Environmental Contamination and Toxicology* (1996) 56: 873-880.

Fareed M, Kesavachandran CN, Pathak MK, Bihari V, Kuddus M, Srivastava AK. Visual disturbances with cholinesterase depletion due to exposure of agricultural pesticides among farm workers. *and Environmental Chemistry* (2012) 94 (8): 1601–1609.

Fedorov LA, Yablokov AL. Pesticides- The chemical weapon that kills life. center for Russian environmental policy. - Moscow (2004).

Fenske RA, Day Jr EW. Assessment of exposure for pesticide handlers in agricultural, residential and institutional environments.: Franklin CA, Worgan JP (eds) Occupational and Residential Exposure Assessment for Pesticides., Chichester, (2005) 13–43

Fishel FM, Nesheim ON, Mark AM. *Toxicity of Pesticides.*, FL: College of Agricultural Science, University of Florida IFAS Extension (2011). available at http://edis.ifas.ufl.edu/pdffiles/PI/PI00800.pdf.

Fogg P, Boxall ABA, Walker A, Jukes AA. degradation in a 'biobed' composting substrate. *Management Science* (2003a) 59:527–537.

Fogg P, Carter A. D.: The development and evaluation of a biological system for pesticide waste and washings. *Symposium* (1998) 70:49–58.

Fogg, P., Boxall, A. B. A., Walker, A. and Jukes, A. Degradation and leaching potential of pesticides in biobed systems. *Management Science* (2003b) 60:645–654.

Franco R, Li S, Rodriguez H, Burns M, Panayiotidis M. Molecular mechanisms of pesticide neurotoxicity. *Interactions* (2010) 188: 289-300.

Freire C, Koifman RJ, Koifman S. Hematological and hepatic alterations in brazilian population heavily exposed to organochlorine pesticides. *of Toxicology and Environmental Health - A* (2015) 78:534–48.

Gangemi S, Gofita E, Costa C, Teodoro M, Briguglio G, Nikitovic D, Tzanakakis G, Tsatsakis AM, Wilks MF, Spandidos DA, Fenga C. Occupational and environmental exposure to pesticides and cytokine pathways in chronic diseases (Review). *Journal of Molecular Medicine* (2016) 38:1012–1020, https://doi. org/10.3892/ijmm.2016. 2728.

Garg UK, Pal AK, Jha GJ, Jadhao SB. Haemato-biochemical and immunopathophysiological effects of chronic toxicity with synthetic pyrethroid, organophosphate and chlorinated pesticides. *Immunopharmacology* (2004) 4: 1709-1722.

Gildon RC, Huffling K, Sattler B, Pesticides and health risks. *of Obstetric, Gynecologic, & Neonatal Nursing* (2010) 39(1):103–110.

Graham M. Pesticides: health, safety and the environment. Wiley & Sons, 2015.

Han R, Wu Z, Huang Z, Man X, Teng L, Wang T, Liu P, Wang W, Zhao X, Hao J, et al., Tracking Pesticide Exposure to Operating Workers for Risk Assessment in Seed Coating with Tebuconazole and Carbofuran. *Management Science* (2021) 77:2820–2825

Harris DA, Johnson KS, Ogilvy JME. A system for the treatment of waste water from agrochemical production and field use. *Crop Protection Conference – Weeds* (1991) 715–722.

Henriksen VV, Helweg A, Spliid NH, Felding G, Stenvang L. Capacity of model biobeds to retain and degrade mecoprop and isoproturon. *Management Science* (2003) 59:1076–1082.

Hock WK, Lorenz ES. *of Pesticides*. Park, PA: The College of Agricultural Sciences, The Pennsylvania State University (2006). available at http://pubs.cas.psu.edu/FreePubs/pdfs/uo222.pdf,https://doi.org/10.1016/j.scitotenv.2017.03.053.

Igbedioh SO. of agricultural pesticides on humans, animals, and higher plants in developing countries. *of Environmental and Occupational Health* (1991) 46(4):218-224.

Iwafune T. Studies on the behavior and ecotoxicity of pesticides and their transformation products in a river. *of Pesticide Sciences* (2018) 43 297–304, https://doi. org/10.1584/jpestics.J18-01.

Jaga K, Dharmani C. Ocular toxicity from pesticide exposure: a recent review. *Health and Preventive Medicine* (2006) 11(3): 102–107.

Jokanović M, Stojiljković MP. Current understanding of the application of pyridinium oximes as cholinesterase reactivators in treatment of organophosphate poisoning. *journal of pharmacology* (2006) 553(1-3):10-7.

Kalyabina VP, Esimbekova EN, Kopylova KV, Kratasyuk VA. Pesticides: formulants, distribution pathways and effects on human health–a review. *reports* (2021) 8:1179-1192.

Karami-Mohajeri S, Abdollahi M. Toxic influence of organophosphate, carbamate, and organochlorine pesticides on cellular metabolism of lipids, proteins, and carbohydrates: a systematic review. *and Experimental Toxicology* (2011) 30(9):1119–40.

Kim KH, Kabir E, Jahan SA. Exposure to pesticides and the associated human health effects. *of total environment* (2017) 575:525–535, https://doi.org/10.1016/j.scitotenv.2016.09.009.

Krieger RI, Bernard CE, Dinoff TM, Fell L, Osimitz TG, Ross JH, Thongsinthusak T. Biomonitoring and Whole Body Cotton Dosimetry to Estimate Potential Human Dermal Exposure to Semivolatile Chemicals. *of exposure analysis and environmental epidemiology* (2000) 10:50–57.

Kumar N, Pathera AK, Saini P, Kumar M. Harmful effects of pesticides on human health. *Annals of Agri-Bio Research* (2012) *17*(2):125-127.

Kumar V, Kumar P. Pesticides in agriculture and environment: Impacts on human health.: Contaminants in Agriculture and Environment: Health Risks and Remediation. and pollution laboratory, Department of zoology and environment science, Gurukula Kangri Vishwavidyalaya, Haridwar, Uttarakhand, India. Environ Media, Agriculture and Environmental Science Academy, Haridwar, India (2019).

Kwong LH, Ercumen A, Pickering AJ, Unicomb L, Davis J, Leckie JO, Luby SP. Soil Ingestion Among Young Children in Rural Bangladesh. *of Exposure Science and Environmental Epidemiology* (2021) 31:82–93.

Lei BL, Zhang KQ, An J, Zhang XY, Yu YX. Health Risk Assessment of Multiple Contaminants Due to Consumption of Animal-Based Foods Available in The Markets of Shanghai, China. *Science and Pollution Research* (2015) 22:4434–4446.

Lemaire G, Terouanne B, Mauvais P, Michel S, Rahman R. Effect of organo-chlorine pesticides on human androgen receptor activation *in vitro. and Applied Pharmacology* (2004) 196:235–46. doi:10.1016/j.taap.2003.12.011.

Macfarlane E, Carey R, Keegel T, El-Zacmay S, Fritschi L. Dermal exposure associated with occupational end use of pesticides and the role of protective measures. *Health Work* (2013) 4:136–141.

Macharia I. Pesticides and health in vegetable production in Kenya. *research international* (2015).

Mergler D. Manifestations of acute and early chronic poisoning. *Encyclopaedia of Occupational Health and Safety* (1998) *1*:7-13.

Mostafalou S, Abdollahi M. Pesticides and human chronic diseases: evidences, mechanisms, and perspectives. *Toxicology and applied pharmacology* (2013) *268*(2):157-177.

Msibi SS, Chen CY, Chang CP, Chen CJ, Chiang SY, Wu KY. Pesticide Inhalation Exposure from Multiple Spraying Sources Amongst Applicators in Eswatini, Southern Africa. *Management Science* (2021) 77:4303–4312.

Narayanasamy P. Postharvest Pathogens and Disease Management. Wiley and Sons, New York (2006). https://doi.org/10.1002/0471751987.

Navarrete-Meneses MDP, P´erez-Vera P. Pyrethroid pesticide exposure and hematological cancer: Epidemiological, biological and molecular evidence. *Reviews on Environmental Health* (2019) 34:197–210, https://doi.org/10.1515/reveh-2018-0070.

Nesheim O, Fishel NFM, Mossler M. Toxicity of pesticides. UF/ IFAS EDIS Document PI-13 (2012). Details available at http://edis.ifas.ufl.edu/pi008, last accessed on 25 April 2014

Ngowi AVF, Maeda DN, Partanen TJ, Sanga MP, Mbise G. Acute health effects of organophosphorus pesticides in Tanzanian small-scale coffee growers. *Journal of Exposure Analysis and Environmental Epidemiology* (2001b) 11:335–339.

Ngowi AVF, Maeda DN, Partanen TJ. Assessment of the ability of health care providers to treat and prevent adverse health effects of pesticides in agricultural areas of Tanzania. *International Journal of Occupational Medicine and Environmental Health* (2001a) 14:349–356.

Ostrea EM, Bielwski MD, Posecion NC, Corrion M. Combined analysis of prenatal and neonatal matrices to detect fetal exposure to environmental pesticides. *Environmental Research* (2009) 116-122.

Page A, Liu S, Gunnell D, Astell-Burt T, Feng X, Wang L, Zhou M. Suicide by pesticide poisoning remains a priority for suicide prevention in China: analysis of national mortality trends 2006–2013. *Journal of affective disorders* (2017) 208:418–423, https://doi.org/10.1016/j.jad.2016.10.047.

Prasad TV. Handbook of Entomology. New Vishal Publication, 2016.

Quijano L, Yus`a V, Font G, Pardo O. Chronic cumulative risk assessment of the exposure to organophosphorus, carbamate and pyrethroid and pyrethrin pesticides through fruit and vegetables consumption in the region of Valencia (Spain). *Food and Chemical Toxicology* (2016) 89:39–46, https://doi.org/10.1016/j.fct.2016.01.004.

Ramchandra AM, Chacko B, Victor PJ. Pyrethroid Poisoning. *Indian Journal of Critical Care Medicine* (2019) 23(Suppl4): S267-S271. doi: 10.5005/jp-journals-10071-23304.

Ramwell CT, Johnson PD, Boxall ABA, Rimmer DA. Pesticide residues on the external surfaces of fi eld-crop sprayers: Environmental impact. *Pest Management Science* (2004) 60:795–802.

Rani L, Thapa K, Kanojia N, Sharma N, Singh S, Grewal AS, Srivastav AL, Kaushal J. An extensive review on the consequences of chemical pesticides on human health and environment. *Journal of Cleaner Production* (2021) 283:124657.

Raschke AM, Burger AEC. Risk assessment as a management tool used to assess the effect of pesticide use in an irrigation system, situated in a semi-desert region. *Archives of Environmental Contamination and Toxicology* (1997) 32:42–49, https://doi.org/ 10.1007/s002449900153.

Schreinemachers P, Chen HP, Nguyen TTL, Buntong B, Bouapao L, Gautam S, Le TN, Pinn T, Vilayson P, Srinivasan R. Too much to handle? Pesticide dependence of smallholder vegetable farmers in Southeast Asia. *Science of the Total Environment* (2017). 593:470-477. https://doi.org/10.1016/j.scitotenv.2017.03.181.

Shah R. Pesticides and human health. In: *Emerging contaminants* (2020) 1-22.

Sharma A, Rai P, Raj D, Anghore D, Sharma P. Antidotes for pesticide poisoning: A review. *Journal of Pharmacy and Pharmacology* (2015) 1:1-4.

Shi JC, Li YL, Liang H, Zheng GJ, Wu YL, Liu WH. Ocps and Pcbs in Marine Edible Fish and Human Health Risk Assessment in The Eastern Guangdong, China. *Archives of Environmental Contamination and Toxicology* (2013) 64:632–642.

Singh NS, Sharma R, Parween T, Patanjali PK. Pesticide contamination and human health risk factor. Modern age environmental problems and their remediation. (2018) 49-68.

Siriwat S, Nganchamung T, Ponrachom C, Siriwong W, Robson MG. Health Risk Assessment of Dermal Exposure to Chlorpyrifos Among Children in Agricultural Areas in Sakon Nakhon Province, Thailand. *Human and Ecological Risk Assessment* (2021) 27:2277–2287.

Surgan M, Condon M, Cox C. Pesticide risk indicators: unidentified inert ingredients compromise their integrity and utility. *Environmental management* (2010) 45:834–841.

Thiermann H, Worek F, Kehe K. Limitations and challenges in treatment of acute chemical warfare agent poisoning. *Chemico-Biological Interactions* (2013) 206:435e443.

Tiemann U. *In vivo* and *in vitro* effects of the organochlorine pesticides DDT, TCPM, methoxychlor, and lindane on the female reproductive tract of mammals: a review. *Reproductive Toxicology* (2008) 25:316–26. doi:10.1016/j.reprotox.2008.03.002.

U.S. Environmental Protection Agency (USEPA). Pesticides: Health and Safety. National Assessment of the Worker Protection Workshop #3. (2007). (Available at:) http://www2.epa.gov/pesticide-worker-safety.

USEPA. Assessing human health risk from pesticide. https://www.epa.gov/pesticide-science-and-assessing-pesticide-risks/assessing-human-health-risk-pesticides (2016). Accessed 18 Dec 2016.

USEPA. Lethal Dosage (LD50) Values. Environmental Protection Agency, United States (2012). Details available at http://www.epa.gov/oecaagct/ag101/ pestlethal.html., last accessed on 5 November 2012.

USEPA. Using toxicity tests in ecological risk assessment, Publication 9345.0-05I, ECO update (1994). Details available at http://www.epa.gov/oswer/riskassessment/ecoup/pdf/v2no1.pdf, last accessed on 2 March 2012.

Valenzuela MC, Arroyo GS, Pietrini VR. et al., Genotoxic biomonitoring of agricultural workers exposed to pesticides in the north of Sinaloa State, Mexico. *Environment international* (2009) 35:1155–1159.

Wang W, Jin J, He R, Gong H. Gender differences in pesticide use knowledge, risk awareness and practices in Chinese farmers. *Science of Total Environment* (2017) 590-591:22-28. https://doi.org/10.1016/j.scitotenv.2017.03.053.

Wang X, Murison J, Wang J, Leong G, Wu, ZC, Li QX. Dermal Exposure Assessment to Trinexapac-Ethyl: A Case Study of Workers in Golf Course in Hawaii, USA. *Environmental Science and Pollution Research* (2021) 28:1072–1076.

Winkler DA. "Overview of Quantitative Structure–Activity Relationships (QSAR)." *Molecular Analysis and Genome Discovery* (2004) 347-367.

Wolf TM, Gallender KS, Downer RA, Hall FR, Fraley FW, Pompeo MP. (1999) Contribution of aerosols generated during mixing and loading of pesticides to operator inhalation exposure. *Toxicology Letters* (1999) 105:31–38.

Worek F, Thiermann H, Wille T. Oximes in organophosphate poisoning: 60 years of hope and despair. *Chemico-biological interactions*, (2016) *259*:93-98.

Worek F, Thiermann H. The value of novel oximes for treatment of poisoning by organophosphorus compounds. *Pharmacology and Therapeutics* (2013) 139:249e259.

World Health Organisation. Public health impact of pesticides used in agriculture. (1990) 128p.

Yohannes YB, Ikenaka Y, Saengtienchai A, Watanabe KP, Nakayam SMM, Ishizuka M. Concentrations and Human Health Risk Assessment of Organochlorine Pesticides in Edible Fish Species from A Rift Valley Lake-Lake Ziway, Ethiopia. *Ecotoxicology and Environmental Safety* (2014) 106:95–101.

Yoshida T, Mimura M, Sakon N. Estimating Household Exposure to Pyrethroids and The Relative Contribution of Inhalation Pathway in A Sample of Japanese Children. *Environmental Science and Pollution Research* (2021) 28:19310–19324.

Yousaf BG, Wang Imtiaz M, Zia-ur-Rehman M, Munir MAM, Niu Z. Bioavailability evaluation, uptake of heavy metals and potential health risks via dietary exposure in urban-industrial areas. *Environment Science and Pollution Research* (2016) 23:22443–22453, https://doi.org/10.1007/s11356-016-7449-8.

Chapter 7

Ecological Impact of Insecticides

Asish Kumar Rout[1,*]
V. Anithaa[2,†]
Anup Chandra[3,‡]
G. K. Sujayanand[3,‖]
Mohammed Muzeruddin Baig[4,¶]
and Rishikesh Kumar[3,#]
[1]ICAR-Indian Institute of Pulses Research, Regional centre, Khordha, Odisha, India
[2]Department of Agricultural Entomology, University of Agricultural Sciences, Dharwad, Karnataka, India
[3]ICAR-Indian Institute of Pulses Research, Kanpur, Uttar Pradesh, India
[4]Central Tasar Research and Training Institute, Piska Nagri, Ranchi, Jharkhand, India

Abstract

Insecticides are often considered quick, easy-to-use, and inexpensive solutions for controlling insect pests in various ecosystems. However, their widespread use comes at a tremendous risk to all living beings, including nontarget organisms. Almost every part of our environment has been contaminated with insecticides. The remnants and residues of the applied insecticides have been detected in soil, air, surface water, and groundwater. Pollinators like honeybees, syrphids, and bumblebees are

* Corresponding Author's Email: asishrout01@gmail.com
† Corresponding Author's Email: anithvel@gmail.com
‡ Corresponding Author's Email: anup.ento@gmail.com
‖ Corresponding Author's Email: sujay2020@gmail.com
¶ Corresponding Author's Email: muzeerbaig89@gmail.com
# Corresponding Author's Email: rishiiari2011@gmail.com

In: Insecticides
Editors: Archana Anokhe and Govindaraj Kamalam Dinesh
ISBN: 979-8-90134-022-6

often killed by direct contact with insecticides while foraging on flowers, and some bees also contaminate their hives while transporting food to the nest. Systemic insecticides usually harm beneficial insects by contaminating the food chain. The honeydew excreted by hemipterans is frequently contaminated with systemic insecticides, which are found to be toxic to beneficial organisms. Apart from mortality, exposure to neonicotinoid insecticides causes various sublethal impacts on aquatic life, including feeding inhibition, reduced body size and body weight, reduced fecundity, and impaired movement in fish, mayflies, and other aquatic insects. Insecticides such as pyrethroids and diazinon impair water quality and are toxic to the *amphipod Hyalella and the freshwater crustacean Ceriodaphnia*. This sets off the entire food web of the marine system. Insecticides affect ecosystem functioning by altering community structure, function, and abundance, as well as reducing primary productivity. Chemical pest control practices, along with other biological methods, minimal use of pesticides, and pest-resistant varieties, as well as disruption of pest populations through sterile insect techniques and bio-remediation methods, have a long way to go to reduce the pesticide load in the ecosystem.

Keywords: pollinators, systemic insecticides, food web, neonicotinoids, bio-remediation

Introduction

Ecology is the study of complex interactions between interdependent creatures living in the same geographic region and interacting with their surroundings at the community level (Johnson & Strinchcombe, 2007). On an individual level, it encompasses the relationships an individual has with various physical and biological elements. The natural surroundings, combined with the living things (biota) that inhabit a specific area, comprise an ecosystem. A farm pond, a rainforest, and a highland meadow are examples of ecosystems. The ecosystem undergoes a specific sequence of events to reach the climax stage. Along with the growth and reproduction of organisms, another important event is the interaction among species. From these various processes and interactions, the structure and function of the ecosystem are determined, as well as the exchange and recycling of matter and energy within it. After passing through various serial stages, the most suitable species dominate, and this is known as the climax species.

An individual of any species is the basic unit of an ecosystem. Individuals interact with numerous biotic and abiotic factors. The following hierarchy of organization is population. They are individuals of the same species. Population ecology is the study of ecology that concerns populations, specifically the interactions between individuals and their environment. The following organization is the community. They are composed of populations of different organisms that interact with one another. There is a phenomenon of co-evolution among the populations of a community. Members of the community share resources with one another. The ecosystem is the next tier of organization. An ecosystem comprises different groups of organisms within a geographically defined area. An ecosystem of a forest, for instance, is made up of plant and animal communities. (A paragraph marked in red is not required).

Insecticides (Pesticides): The Barrage of Biocides

Chemical substances known as pesticides are used to control weeds, fungi, rodents, and insects. These can be plant growth regulators, insecticides, fungicides, molluscicides, herbicides, nematicides, and other substances. Pesticides play important roles in various commercial and food-based industrial processes, including aquaculture, agriculture, food processing, and storage. They are typically employed to prevent infections caused by vectors (Sharma et al., 2019). Pests are any living thing that poses a risk to people or any other organisms of human interest. Pesticides are chemicals used to eradicate pests or stop them from growing.

Pest management has employed a range of chemicals since antiquity. One well-known example of such a pesticide for insect and mite control is the sulphur compounds. For more than 2,000 years, pyrethrum, a pesticide derived from the plant *Chrysanthemum cinerariaefolium*, has been used for pest control (Unsworth, 2010). Before the introduction of dichlorodiphenyltrichloroethane (DDT) as a potent pesticide in 1939 by Paul Hermann Müller, both organic and inorganic chemical compounds were frequently used to reduce insect populations (Abubakar et al., 2020). Consequently, the need for DDT grew on a global scale, paving the way for the synthesis of novel chemicals with pesticidal properties. Subsequently, organophosphates (OPs) and carbamates (CMs) replaced DDT during the third quarter of the 20th century. The total quantity of pesticides used worldwide in 2020 was estimated to be approximately 4.5 million metric tons, with China consuming the largest share (1.86 million metric tons), followed by the United States and Brazil. According to WHO data, pesticide use has been rising annually throughout the Southeast Asia region. Developing nations, such as Cambodia, Laos, and Vietnam, account for 20% of the region's pesticide consumption.

One of the leading producers of pesticides in Asia, India produces 90 thousand tons of organochlorine pesticides annually, including DDT and benzene hexachloride. The country with the highest average pesticide usage from 2010 to 2014 was Japan (18.94 kg ha^{-1}), followed by China (10.45 kg ha^{-1}) and Mexico (7.87 kg ha^{-1}). Pesticide usage in India is lower compared to developed countries. (0.26 kg ha^{-1}) (Khan et al., 2010). Insecticides account for 29.5% of pesticide contributions, while herbicides contribute 47.5%. Many factors are taken into consideration while classifying insecticides. The method of entrance, chemical composition, and target that the pesticide kills are the most widely utilized classification criteria. However, the World Health Organization (WHO) prioritizes public health when classifying pesticides based on their toxicity and adverse effects.

The principal benefits of pesticides are the anticipated immediate gains upon treatment, such as the eradication of pests, which primarily increase the yield and quality of crops. The productivity and production of most countries have increased significantly. Insect pests can significantly reduce the production of harvestable crops. Hence, insecticides have been essential to the production and processing of food. Agricultural productivity is increased, and food is preserved through the use of pesticides, notwithstanding the associated risks. By using it sparingly and utilizing multiple pesticide categories, the risk of overuse, exposure, and adverse effects can be reduced (World Health Organization, 2009). The regular functioning of an ecosystem can be negatively impacted by insecticides discharged into the environment in various ways, ranging from long-term repercussions to transient modifications. Even while utilizing pesticides in agricultural and public health has many benefits, there are typically negative impacts on the environment and public health when they are used. Due to their high biological toxicity (both acute and chronic), insecticides hold a unique position among environmental pollutants. By definition, chemical insecticides are poisonous. All life forms other than the pest species that the pesticide is intended to kill can typically be harmed by it. Because of their characteristic, they are best characterized as biocides, which can eradicate all living organisms, even if the method of action of certain pesticides is said to be selective. Insecticide use has been shown to have numerous negative impacts; hence, solving this problem requires efficient waste management techniques.

Classification of Insecticides

Based on their specific chemical compositions and amounts, insecticides exhibit varying degrees of toxicity in biological systems. According to Garcia et al., (2012), pesticides can be categorized into two main groups according to their chemical and physical characteristics. Insecticides can be classified based on the type of pest they target and their origin (natural and synthetic). The bulk of insecticides is composed of organic compounds, which are divided into Four Main Categories: Organochlorines, organophosphates, pyrethroids, and carbamates. Biopesticides, also referred to as naturally occurring pesticides, are produced by living organisms, including fungi, bacteria, and plants (Bhatt et al., 2021). The two main variables that determine the acute and chronic toxicity of insecticides are the dosage (amount) and exposure duration (time). The term "acute toxicity" refers to the toxicity of a pesticide to humans, animals, and plants following a specific, brief exposure. Even in small doses, an insecticide with a high acute toxicity can be lethal to humans. The lethal dosage (LD_{50}) divides acute toxicity into two categories: acute cutaneous (dermal) toxicity (e.g., extremely toxic: 50 mg/kg body weight of rat; highly toxic: 50-200 mg/kg body weight of rat; moderately toxic: 200-2,000 mg/kg body weight of rat) and acute oral toxicity (e.g., extremely toxic: < 5-mg/kg body weight of rat body weight; extremely toxic: 5–50 mg/kg; moderately toxic: 50–2,000 mg/kg) Not required in text as it is already mentioned in table (Table 1). However, the World Health Organization (WHO) only recognizes acute toxicity for pesticide categorization. Chronic toxicity refers to the lethal effects of insecticide exposure that persist for an extended period. Individuals who work with pesticides for an extended period are concerned about the chronic toxicity of pesticides. Nowadays, pesticides are categorized into "WHO Hazard classifications." This categorization was combined with the "Globally Harmonized System (GHS) Acute Toxicity Hazard Category" (Table 2).

An additional classification scheme is predicated on the mechanism of admission, which is subdivided into five distinct types.

1) Systemic insecticides: these insecticides are absorbed by plants or animals and travel through the vascular system to other locations,

such as inside plants, where they enter untreated tissues of roots, stems, or leaves.

2) Non-systemic or contact insecticides: these require physical contact with the pest to work;
3) Stomach poisons: these enter the digestive tract and are absorbed inside the insect's body; they are used to control vectors, such as mosquitoes or black flies.
4) Fumigants: these pesticides are used as poisonous gases or vapour that enter the pest's respiratory system via spiracles and kill it; and
5) Repellents: these pesticides are used to keep pests away from treated objects (Yadav and Devi, 2017).

Table 1. Classification of pesticides according to WHO guidelines (World Health Organization, 2009)

Class	LD_{50} of a rat		Hazard level
	Dermal (mg/kg body weight)	Oral (mg/kg body weight)	
Ia	<50	<5	Extreme
Ib	50–200	5–50	Highly
II	200–2000	50–2000	Moderately
III	>2000	>2000	Slightly
U	≥5000	≥5000	No acute hazard

Table 2. Classification of pesticides according to the Globally Harmonized System GHS (World Health Organization, 2009)

Category	Classification criteria			
	LD_{50} (mg/kg body weight) of the rat (dermal)	Hazard description	LD_{50} (mg/kg body weight) of the rat (oral)	Hazard description
1	<50	Lethal if it comes into skin contact	<5	Lethal if consumed
2	50–200	Lethal if it comes into skin contact	5–50	Lethal if consumed
3	200–1000	Toxic if it comes into skin contact	50–300	Toxic if consumed
4	1000–2000	Harmful if it comes into skin contact	300–2000	Harmful if consumed
5	2000–5000	Possibly harmful if it comes into skin contact	2000–5000	Possibly harmful if consumed

Transformation and Bio-Accumulation of Insecticides After Their Application

This chapter offers a more detailed examination of the primary ecological impacts of insecticides. It is based on the available information about how pesticides behave and the harm they can cause to ecosystems and the environment when used excessively or carelessly. The consequences could range from a minor alteration to the ecosystem's regular operation to the extinction of species. Since the creatures in the ecosystem are entangled in intricate, interdependent relationships, the extinction of a major species could cause the ecosystem as a whole to collapse. The present stringent laws against the use of pesticides in the courts are mainly due to these effects.

The study of pesticide transport, change, and end fate in different environmental compartments is known as the chemo-dynamics of pesticides. The four main components of the environment — soil, water, air, and biota — are responsible for the transformation of insecticides. There are numerous potential sources of insecticides in the environment due to their extensive use and disposal by farmers, industries, and the general public. After being discharged into the environment, insecticides can have a variety of outcomes. Spray-applied insecticides have the potential to travel through the atmosphere. When pesticides are sprayed directly on the soil, they have the potential to seep through to lower soil layers and groundwater or run off into adjacent bodies of surface water (Harrison, 1990). The extensive potential of such movement has resulted in the global spread of DDT and the presence of pesticides in water bodies, such as the Great Lakes, that are distant from their core use areas. Pesticide behaviour and biological activity in soil are determined by physical and chemical qualities, including molecular weight, lipophilicity, and volatility (Beulke et al., 2004).

Generally speaking, the fate of pesticides in an ecosystem within the soil is determined by both biological and abiotic transformations, which are influenced by the presence of living organisms and the pesticide's physicochemical properties. The transport, adsorption, and degradation of pesticides are key processes that regulate their behavior and fate in soil. These activities are further divided into leaching, runoff, microbiological degradation, and soil and plant absorption based on the pesticide's mode of movement. There are different pesticide conversions in the soil system. Adsorption processes are driven by chemical or physical forces, such as van der Waals forces (Gevao et al., 2000). Pesticides can disperse into the atmosphere by runoff or by diffusion, which can subsequently contaminate aquatic bodies. The adsorption of pesticides by both target and nontarget organisms is primarily caused by the physical and chemical characteristics of the soil and pesticides, in conjunction with other environmental factors. This process is referred to as bioaccumulation. Pesticides can leach up to the groundwater table, leaving water vulnerable to pollution.

When pesticides leach into groundwater in sufficient quantities, they can pose a hazard to both human and animal health. The leaching process may be influenced by the physico-chemical and biological characteristics of the soil when pesticides are used for agricultural purposes. Pesticides are translocated into soil and water as a result of various agricultural practices, which depend on several factors, including pH, temperature, moisture, organic matter content, and soil texture. Soil persistence can last for weeks, months, or even years. Furthermore, the mobility and persistence of chemical compounds influence how easily they leach and seep, thereby increasing the risk of water pollution (Tudi et al., 2021).

Ecological Impact of Insecticides

Impact on Natural Systems

The primary goal of applying pesticides in fields and the environment is to control crop pests and disease vectors. To boost agricultural yields and enhance public health, this has been a conscious human endeavour. Some pesticides sprayed in the environment have been shown to have long-term residual effects on the environment. In contrast, others, when improperly handled, can have immediate and deadly impacts. For example, it has been demonstrated that organochlorine pesticides are persistent in the environment. As a result, they can contaminate food products, soil, groundwater, and air and produce health risks when they come into direct contact with humans. Human exposure to pesticides is the primary cause of several ailments, including reproductive issues, endocrine disturbance, cancer, respiratory conditions, and skin ailments. Environmental scientists are concerned about the environmental impact of pesticides. They are working to understand their behavior in the environment in order to develop a viable replacement that protects people from their harmful effects. The world's ecosystems remain under constant threat from pesticides, even fifty years (almost a century) after Rachel Carson warned everyone in an elegant essay about the devastating effects pesticides have on birds and beneficial insects. Every year, a significant chemical attack on our ecosystem begins. This toxic assault infuriates other people. Various suburbanization efforts and dammed rivers are putting strain on our ecosystems, endangering the lives of numerous birds, fish, insects, and small aquatic animals. That the food web is built upon. Pesticides, in general, lessen species variety in the realm of animals and aid in the reduction of animal and plant populations through habitat destruction, decreasing food supply, and hindering reproduction. A mixture of semi-volatile pesticides and aerosol particles adsorbs them. Gas-phase reactivity determines the half-lives of these particles, which range from a few days to several months. Oxidation and photochemical reactions cause persistent organic pollutants (POPs), which are found in the air, to transform from their naturally occurring forms into highly hazardous ones. Based on Woodrow et al., (2018), the low solubility of these POPs in water, as well as climate factors such as temperature and humidity, influences their migration. In contrast to former organochlorine pesticides, current-use pesticides (CUPs) are less harmful, more biodegradable, and less persistent in the environment.

Impact on Terrestrial Systems

Impact on soil systems: Although pesticides are primarily used to protect crops, they can also pollute the soil in several ways. Excessive levels of insecticides that enter water bodies and runoff are broken down and adsorbed as they travel through the soil. Pesticides damage beneficial biota, including microbes, honey bees, predators, birds, plants, and small animals, when they are applied excessively and arbitrarily to a variety of crop species (Pokhrel et al., 2018).

Impact on soil nutrient cycling: A significant portion of pesticides used in the environment end up in the soil, where they facilitate soil-building processes and the cycling of nutrients back into living plants. By directly or indirectly affecting the soil organisms involved in these processes, pesticides can interfere with the ecosystem's natural nutrient cycling. Pesticides in soil inhibit nitrogen fixation, a process essential for the growth of higher plants. There is evidence that the insecticides DDT, methyl parathion, and pentachlorophenol, in particular, disrupt the chemical communication between legumes and rhizobium. Crop yields are affected when this mutualistic chemical signalling is diminished because less nitrogen is fixed. The production of root nodules in these plants saves the global economy \$10 billion annually in synthetic nitrogen fertilizer. Reduced soil fertility and productivity result from any interference with the ecosystem's natural nutrient cycling by pesticides or other pollution sources (Fox, 2007).

Impact on fertility and productivity of soil: Numerous compounds found in pesticides are long-lasting soil pollutants that have a negative impact on soil conservation, potentially lasting for decades. Pesticide usage reduces the soil's overall biodiversity. Higher soil quality is achieved by avoiding the use of chemicals, which also has the added benefit of increasing soil organic matter, thereby enhancing water retention. In drought years, when organic farms have yielded 20–40% more than their conventional counterparts, this helps farms increase their yields. Because organic matter binds to and aids in the breakdown of pesticides, a lower amount of organic matter in the soil increases the amount of pesticide that will exit the application area. Soil-applied pesticides, for instance, can lessen the amount of vegetation covering the ground, which encourages runoff and wind-induced soil erosion. Because soil erosion alters the soil's structure, there is an uneven distribution of soil fertility. Plants cannot flourish on barren terrain with inadequate soil fertility and soil structure. This land's ecological limitations prevent it from supporting other living forms, which could lead to the collapse of the ecosystem (Lotter et al., 2003).

Impact on soil and associated flora: Soil is a complex mixture of minerals, organic matter, water, gases, and living organisms. In other words, the soil system acts as a micro-ecosystem on its own. The flora and fauna thriving within it play a crucial role in cycling nutrients, sequestering carbon, maintaining structure, biocontrolling, and ensuring its health. It is necessary for there to be diversity and heterogeneity in the flora and fauna to sustain agricultural productivity and maintain the ecological resilience of the ecosystem. Soil microorganisms play a key role in carbon and nitrogen cycles in animals, plants, agriculture, and the food web (Cavicchioli et al., 2019). Soil fauna play a crucial role in nutrient cycling by utilizing organic matter, accelerating the decomposition of detritus through the distribution of nutrients via feeding, fragmentation, or microbial transport (David C. Coleman, 2008) The fate of insecticides in the soil, in general, and their impact on microbes and their activity depend on the nature of the pesticides used, their water solubility, persistence, rate of adsorption, quantity applied, and soil characteristics, such as sand, clay, and organic matter content (Subhani et al., 2000). The higher percentage of clay and organic matter in the soil provides a greater number of adsorption sites due to increased binding, resulting from a greater number of charge sites, which also decreases the potential for a pesticide to leach and drift. This allows the residues to remain in the soil for more extended periods (Gevao et al., 2000). The persistence of a pesticide is measured by its half-life. The longer the half-life, the greater the persistence of pesticides in the soil system (Wang et al., 2018). The beneficial soil microorganisms and the associated transformations in the soils are adversely affected by the residues of different pesticides. Various chemical pesticides affect the growth and biological properties of both legume plants and rhizobia. Symbiotic nitrogen fixation is strongly inhibited due to the blockage of biochemical pathways and interactions between the legume plants and symbionts (Ahemad and Khan, 2013). The decrease in the abundance of Enterobacter, Aeromonas, Bordetella, and Staphylococcus in soils exposed to pesticide overuse impairs the decomposition of organic matter, plant growth, homeostasis, and the biocontrol of insect pests (Hayward et al., 2010).

Similarly, the reduction of diversity and load of fungi in the soil negatively affects the decomposition of surface litter by 36% (Sigler and Turco, 2002). Soil application rates of some organophosphorus insecticides, such as chlorpyrifos, trichlorfon, and quinalphos, may initially enhance the growth of rhizospheric fungi; however, bacterial populations typically take over within 50-60 days (Pandey and Singh, 2004). Soils contaminated with copper and arsenic compounds take many years for the regeneration of fungi (Zwieten et al., 2004), and it ultimately reduces the microbial degradation of DDT in soil (Gaw et al., 2003). Protozoans are also equally sensitive as the other soil microbes. It has been reported that the number of ciliates has decreased slightly. However, testate amoeba species can be reduced by up to 50% in agroecosystems treated with pesticides. Cry toxins from *Bacillus thuringiensis* have minimal deleterious effects on microbes, including nematodes and protozoans. However, some effects of transgenic plants on the soil microflora have been reported, which are mostly due to various transient factors and not to the presence of Bt toxins (Icoz and Stotzky, 2008).

Impact on soil macrofauna: Animals that are one centimetre or more in length and smaller than an earthworm can be categorized under macrofauna. These can include centipedes, millipedes, myriapods, snails, slugs, spiders, and adult and immature arthropods, among others. These have both direct and indirect effects on trash burning and the cycling of nutrients. Detritivores are known for their low assimilation, ranging from 10 to 30%. Insecticide applications on soil and crops result in a reduction of the soil macrofauna, which in turn disrupts the complex soil structure. Deleterious effects of organo-phosphorus and carbamate insecticides have been observed on the soil macrofauna. Millipedes are more tolerant of broad-spectrum insecticides, but their populations declined over the years in cabbage plots treated with DDT because residues of this persistent OC accumulated in their bodies, thus reducing their reproductive capacities (Brown, 1978). In general, soil macrofauna communities exhibit more sensitivity to a range of pesticides than meso- and microfauna due to longer generation times and lower population densities (Liess & von der Ohe, 2005). Similarly, soil fauna without an exoskeleton (such as earthworms, nematodes, and enchytraeids) are more sensitive than arthropods with an exoskeleton, as they may act as a protective barrier against various pesticides (Jänsch et al., 2006). The application of recommended doses of pyrethroids has a severe impact on the population of predatory mites, as they are physiologically susceptible to this insecticide. This, in turn, increases the springtail population in the soil ecosystem. (Badji et al., 2007). However, chlorpyriphos diminishes the Collembolan populations, and it takes at least a year for the species to recover. Similarly, ground spiders were also significantly reduced in grassland ecosystems treated with recommended doses of organophosphorus insecticides. Consequently, *Ceratophysella denticulate*, a springtail, increases and dominates in the community structure due to the absence of predatory mites (Fountain et al., 2007). Aerial application of broad-spectrum insecticides to crops, fruit orchards, or locust breeding sites often decimates the populations of nontarget organisms, such as rove beetles, tiger beetles, predatory ants, dermapterans, centipedes, millipedes, and spiders. Predatory hymenopterans and Tenebrionids are knocked off after the application of fenitrothion in soil in grassland ecosystems. However, their populations can recover within a year (Silva et al., 2019). Predatory ants and other soil-dwelling hymenopterans are more sensitive to sprays of acetylcholinesterase inhibitors and insect growth regulators, such as diflubenzuron. Growth regulators, such as triflumuron, on the other hand, control locusts and have a minimal impact on most arthropods, except for predatory hymenopterans. Hence, incorporating these parasitoids into IPM programs raises concerns. Similarly, spinosad has little effect on most predatory hexapods; however, field studies have shown a decline in reproductive fitness and reduced longevity in several hymenopteran parasitoids (Williams et al., 2003). Most organochlorine insecticides reduce the populations of carabid and staphylinid beetles in horticultural ecosystems. Despite their high toxicity, the effect of acetylcholine esterase inhibitors, such as pyrethroids and imidacloprid, on carabid populations has little impact at recommended doses (Sánchez-Bayo et al., 2011). Residues of the active chemicals applied on the livestock, like cypermethrin, spinosad, avermectins, etc., are usually accumulated in the manure, thus annihilating dipterans and other coleopteran beetles. This leads to a change in the community structure of the saprophytes through alteration in the species richness and diversity (Iwasa et al., 2005). However, insect growth regulators have no considerable deleterious effects on these populations.

Impact on the resurgence of pests: The resurgence of pests is entirely at odds with the purpose of pesticides, which are designed to control insect and other arthropod populations. However, the most well-documented side effect of pesticides is the resurgence of pests. It has long been recognized that an increase in natural enemies can release the pest population from control, resulting in an insecticide-induced resurgence of arthropod pests. The primary reason for the resurgence of pests is the reduction or elimination of natural predators and parasitoids by the use of insecticides. Although they are not the primary cause, additional processes, including improved host-plant nutrition, enhanced pest fecundity due to more productive crop varieties, or increased attractiveness, may also contribute to the revival (Hardin et al., 1995). The wasp parasitoid Aphytis melinus (Hymenoptera: Aphelinidae), which was the natural pest controller of the citrus red scale *Aonidiella aurantii* (Hemiptera: Diaspididae), was likewise eradicated by DDT when it was first used in California. The scale's population doubled in a year and then skyrocketed in the ensuing years. The autumn armyworm, *Spodoptera frugiperda* (Lepidoptera: Noctuidae), is a significant pest of maize. The predatory ant Ectatomma ruidum (Hymenoptera, Formicidae) is effective in eliminating its pupae. However, because the inscticide severely reduced the ants' ability to forage, the number of this pest rose in Nicaragua when corn crops were treated with carbofuran. In a well-known example of pest resurgence, thrips populations increased in English wheat fields after cypermethrin was applied.

Impact on secondary pest outbreaks: The eradication of major crop pests has the unintended consequence of releasing herbivore species from competition, which can lead to increased outbreaks of secondary pests. Since the introduction of DDT in Sudan, Mexico, and eventually Australia in the late 1950s, outbreaks of secondary pests in cotton have gained widespread recognition. The primary reason for these epidemics was the reduction of the natural enemies of secondary pests, including parasitoid wasps and predators, which controlled the populations of secondary pests like *Bemisia tabaci*, the silverleaf whitefly. Understanding this replacement requires an understanding of the pest species' susceptibility to pesticides. Under exposure to fenitrothion, the dominance of the lesser grain borer, *Rhyzopertha dominica* (Coleoptera: Bostrichidae), over the maize weevil, *Sitophilus zeamais* (Coleoptera: Curculionidae), shifted in experimental studies (Cordeiro et al., 2014). This suggests that insecticide-mediated species interactions have indirectly altered population densities, which raises concerns as a potential cause of secondary pest outbreaks. Applying chlorpyrifos to corn crops to control *S. frugiperda* resulted in a considerable decrease in predatory ants' foraging activity. Higher levels of the secondary pest of that crop, the maize leafhopper, *Dalbulus maidis*, were seen as a result of the suppression of both the primary pest and its principal natural control agent. In another instance, the insecticide chlorpyrifos killed the predatory phytoseiid mites, which resulted in increased densities of *Tetranychus sp.* (Acari: Tetranychidae) spider mites in vineyards (Prischmann et al., 2005). The main pests of the maize crop, including the cotton bollworm, *Helicoverpa sp.* (Lepidoptera: Noctuidae), the armyworm, *Spodoptera sp.* (Lepidoptera: Noctuidae), the corn rootworm, *Diabrotica sp.* (Coleoptera: Chrysomelidae), the corn borers, *Ostrinia sp.* (Lepidoptera: Crambidae), and the maize weevil, *S. zeamais,* were eliminated when commercial crop varieties of cotton and corn were replaced with genetically modified crops of Bt-cotton and Bt-corn. While the widespread use of Bt cotton has decreased the amount of insecticides sprayed on cotton fields, it has also led to an increase in the number of stink bug pests, *Nezara viridula* (Hemiptera: Pentatomidae), which are highly damaging. Aphids, whiteflies, stink bugs, and other herbivorous insects have all thrived and become pests in those crops as a result of the GM variety. Simultaneously, the reduction in the density of their previously prevalent host pests may cause a decline in the size and species richness of parasitoids. Similarly, the main rice pests, including stemborers, *Chilo sp.* and *Scirpophaga sp.* (Lepidoptera: Crambidae), and rice leafrollers, *Cnaphalocrocis medinalis* (Lepidoptera: Crambidae), have been suppressed over many years by the use of insecticides in rice crops. These pests have been replaced by brown planthoppers, *Nilaparvata lugens* (Hemiptera: Delphacidae), and rice gundhi bugs, *Leptocorisa sp.* (Hemiptera: Alydidae), which are now the predominant pests.

Impact on the natural enemies of crop pests: When pesticides are sprayed on a crop, target pests and other nontarget insects are often eliminated; however, the time it takes for an individual insect to die might vary, ranging from a few minutes to many days, contingent upon the exposure dose that it receives. Natural enemies that consume the afflicted species in the interim may become secondarily poisoned, lose their ability to hunt, or even perish. For instance, when the spined soldier insect, *Podisus maculiventris* (Hemiptera: Pentatomidae), fed on diamond moths, *Plutella xylostella* (Lepidoptera: Plutellidae), its ability to prey was compromised, and its weight decreased when imidacloprid was administered at the prescribed rates in cabbage plots, but strangely, the pesticide did not considerably lower the number of moths

(Resende-Silva et al., 2019) in both lab and field tests. Secondary poisoning most frequently results in the death of natural enemies. Thus, after feeding on second-instar cotton bollworm larvae, *Helicoverpa armigera* (Lepidoptera: Noctuidae), that had been treated with azadirachtin, the survival of lacewing larvae decreased in laboratory trials. Similarly, after feeding on lettuce aphids, a significant percentage of lacewing larvae perish in the laboratory when treated with different systemic insecticides. In both laboratory and field settings, the survival of the ladybug *Cycloneda sanguinea* (Coleoptera: Coccinellidae) fed on aphids treated with thiamethoxam and imidacloprid was dramatically reduced.

In contrast, residues of dimethoate in prey aphids treated at field exposure rates significantly reduce the carabid predators. Because the larvae feed on hosts laced with pesticides or insect growth regulators, parasitoid wasp populations frequently collapse due to failure of emergence. Pyriproxyfen is an example of the latter, as it only affects the eggs and larvae of insects, rather than the adults. Therefore, the parasitoid wasp is observed ovipositing when exposed to the eggs of the brown stink bug (Hemiptera: Pentatomidae) generated by females treated with pyriproxyfen (Penca and Hodges, 2017). However, adult wasps were unable to emerge from the host eggs. Lufenuron has been shown to decrease the survival of parasitoid wasps by modifying the biochemistry and histology of the midgut epithelium.

Impact on the pollinators: It has been proposed that the continuous and growing use of insecticides, particularly neonicotinoids and fipronil, is responsible for the rise in the incidence of infections and viral illnesses among pollinators in recent decades. Because pesticide residues can be found in treated crops as well as in the flowers of weeds, shrubs, and trees in the surrounding area, bees frequently consume a cocktail of pesticides while grazing on pollen and nectar. The amounts taken can be sublethal in most circumstances. However, they also cause energy-consuming detoxifying procedures that frequently lead to stress. Due to this weakened immune system, the insects are less able to fight off infections, parasites, and diseases. Unsurprisingly, exposure to the neonicotinoid imidacloprid in food caused honey bees, *Apis mellifera* (Hymenoptera: Apidae), to become sublethally infected with the microsporidian gut parasite *Nosema spp*. However, in contrast to some hazardous metals, imidacloprid did not affect the gut microbiota of bumblebees, suggesting that these indirect effects are species- and chemical-specific. Compared to bees not exposed to the systemic insecticide, honey bees that collected pollen and nectar from cornfields treated with thiamethoxam as seed coatings had considerably greater pathogen burdens of the mite *Varroa sp.* The mite significantly decreased colony performance and survival, and impaired the bees. The wax moth, *Vitula edmandsae* (Lepidoptera: Pyralidae), was also more prevalent in colonies of eastern bumblebees, *Bombus impatiens* (Hymenoptera: Apidae), that pollinated blueberries treated with the neonicotinoid acetamiprid and the fungicide propiconazole, than in colonies of the same bumblebees feeding on untreated shrubs or treated with a single pesticide. Insect pollinators are few in pesticide-treated plots, whereas the parasitic Microsporidia parasites of leaf-cutter bees, *Megachile sp.* (Hymenoptera: Megachilidae), were highly prevalent.

Impact on the development of pest resistance: Through natural selection, pests may eventually become highly resistant to the pesticide. Pest resistance to a pesticide is commonly managed through pesticide rotation, which involves alternating among pesticide classes with different modes of action to delay the onset of or mitigate existing pest resistance. Tank mixing pesticides involves combining two or more pesticides with different modes of action in a single environment. Pests may evolve to become resistant to pesticides due to the continued use of pesticides in a particular environment.

Impact on human health: According to Sacramento (2008), there are three ways that pesticides can enter a person's body: orally, through the consumption of contaminated food or water; dermally, through direct skin contact; and through aerosols, dust, or vapors that contain pesticides. Pesticides are sprayed on food, particularly fruits and vegetables; they leak into the ground and soil, contaminating drinking water, and the spray can spread and contaminate the atmosphere. Depending on the chemical's toxicity as well as the duration and intensity of exposure, pesticides have more detrimental impacts on human health (Lorenz, 2009). Farm labourers and their families are most commonly exposed to agricultural pesticides due to their proximity to the chemicals. However, a certain quantity of pesticides is present in every human body, as evidenced by fat samples. Because of their tiny size and underdevelopment, children are the most vulnerable to the effects of pesticides. Over time, these substances may accumulate in the body. According to Miller (2004), the effects of pesticide exposure can vary from minor skin irritation to birth deformities, cancers, genetic alterations, blood and nerve diseases, endocrine disturbance, and even coma or death.

Impact of Insecticides on Aquatic Ecosystems

Aquatic ecosystem pollution by insecticides has become a serious concern. There is a significant risk of pesticide drift and runoff because most rivers pass through productive agricultural areas. According to a review of the literature, the primary contaminants in the water column and biota are hexachlorocyclohexane (HCH), DDT, and endosulfan. In contrast, the primary pollutants in the sediment phase are HCH, DDT, aldrin, and dieldrin. It has been discovered that significant riverine systems, such as the Ganges and its tributaries, such as the Yamuna, are severely contaminated. In tests of potable water, pesticides such as DDT, endosulfan, endrin, aldrin, dieldrin, and heptachlor have been found. According to a case study conducted in the Hooghly River basin in West Bengal, India, insecticides were found to be more common in river water samples than herbicides and fungicides. The majority of river samples exceeded the pesticide standards for drinking water in Europe, and organochlorine pesticides were prevalent in the adjacent agricultural areas of the river. There are several ways that persistent organic pesticides and CUP pesticides get into water bodies. These include atmospheric precipitation, as well as the release of raw chemical waste from chemical or pesticide manufacturing industries into rivers and other water bodies. Once inside these water bodies, the pesticides travel great distances and contaminate aquatic environments, having a detrimental effect on aquatic ecosystems. In aquatic systems, these pesticides build up and spread from lower to higher trophic levels, directly impacting aquatic flora and fauna. Through ingestion or other mechanisms, these pesticides then influence human health (Agarwal et al., 2015).

Impact on crustaceans and hexapods: Arthropods comprise a diverse range of insects that fall into several trophic categories, commonly found in aquatic environments. These insects include scavenger beetles, scrapers, and detritivorous larvae of mayflies (Ephemeroptera), stoneflies (Plecoptera), and caddisflies (Trichoptera), as well as grazers such as chironomids (Diptera). Pesticides are recognized to have direct effects on the species mentioned above (Basley et al., 2018). All aquatic arthropods, especially zooplankton and insect larvae, are highly susceptible to insecticides. Since pesticides evaporate from water quite quickly, there are greater residues of these chemicals in streams near agricultural fields than in rivers and lakes. Systemic insecticides are an exception, as they remain persistent in all of these media and are soluble in water. Aquatic populations are typically affected by pesticides directly, as zooplankton crustaceans and insect larvae are highly susceptible to their effects.

Furthermore, decreases in predatory insects, including water bugs and dragonfly nymphs, usually increase the quantity of their prey, which includes tadpoles and snails, who gain from this indirect effect (Relyea and Hoverman, 2008). Insecticide mixtures can have disastrous consequences for vulnerable zooplankton and benthic arthropod species. For instance, within 24 hours of treatment, the density of the water flea, *Daphnia magna* (Crustacea: Cladocera) and the amphipod *Hyalella azteca* (Crustacea: Amphipoda) was significantly reduced by a combination of permethrin, λ-cyhalothrin, and chlorpyrifos. Although the OPs had an immediate effect, endosulfan caused a lag effect by decreasing the number of amphibians and water fleas. The combination treatment had devastating effects on the entire community, with long-term repercussions for the mass of amphibians and indirect, specific effects on phytoplankton and water quality. Imidacloprid not only directly affects benthic macroinvertebrate assemblage, but it also impairs antipredator behavioural responses in two insect species: the midge, *Chironomus riparius,* and the caddisfly, *Sericostoma vittatum,* when exposed to sublethal concentrations of the insecticide. While the shredder caddisfly *S. vittatum's* leaf breakdown was slowed down by chlorantraniliprole, both these effects were amplified in the presence of golden-ring dragonfly nymphs, *Cordulegaster boltonii* (Rodrigues et al., 2018).

Impact on Fishes

Insecticides may affect the reproductive systems of Atlantic salmon (*Salmo salar*) and brown trout (*Salmo trutta*). Furthermore, other research on fish exposed to the pesticide found a variety of developmental defects. Numerous investigations have revealed that pyrethroids are detrimental to fish reproduction and the early phases of embryogenesis. Fish that are young may exhibit delayed production of the egg proteins (vitellogenin and choriogenin) due to exposure to pyrethroids such as bifenthrin and permethrin. Wu et al., (2020) reported that deltamethrin, a second-generation (type II) pyrethroid neurotoxic insecticide, was found to be detrimental to the development of the swim bladder in zebrafish embryos at doses of 20 and 40 g/L.

Similarly, pesticides can affect the activity of acetylcholinesterase (AChE), which can affect fish's neurological systems and lead to a variety of neurotoxic effects (neurotoxicity). Pesticide exposure affects fish species that have been demonstrated to exhibit altered AChE activity in *Labeo rohita*. Moreover, the *Catla catla* brain exhibited neurotoxicity and apoptosis in response to cypermethrin (CYP). Pesticides also harm fish's endocrine systems. These chemical compounds have the potential to cause molecular toxicity in fish species such as *L. rohita* and *Cirrhinus mrigala* when used in significant quantities (Jindal and Sharma, 2019).

Future Directions and Conclusion

The environmental pollution and the corresponding degradation of terrestrial and aquatic ecosystems resulting from the constant use of pesticides should not be overlooked. We all depend on the ecosystem services that the environment provides for maintaining homeostasis and supplying food. Natural ecosystems are supposed to be resilient, and there must be a balance among and between populations of various organisms. However, due to the constant use of insecticides in crops, forests, pastures, and range lands, water bodies have, over the years, reduced the sustainability of the food system for many vertebrates as well as invertebrates. Thus, the glut in crop productivity we enjoy has a tradeoff with the reduction of aquatic as well as terrestrial communities at a global scale. The use of insecticides in various forms has expanded significantly in both space and time, resulting in various forms of ecological damage. Among different categories of pesticides, organophosphates, carbamates, organochlorines, and pyrethroids are the most predominantly used insecticides, and they pose several deleterious human and environmental concerns due to their low rate of degradation, high stability, and prolonged half-lives. These insecticides can drift and become biomagnified at different trophic levels of the food chain. Insecticide contamination is hazardous to every ecosystem and is undoubtedly harmful to all organisms that depend on it. Approaches for insect-pest management include ecological engineering, application of need-based chemicals, and integrated pest management.

Non-chemical measures, such as cultural, biological, and host plant resistance, should be implemented and scaled up to reduce pesticide load. Cultural control is often attributed to the non-chemical alteration of factors responsible for pest attacks, such as maintaining field sanitation and crop rotation. Control by natural enemies, genetic control through breeding resistant crops (e.g., Hessian fly on wheat), sterile insect techniques (e.g., tsetse fly), biotechnology, and genetic engineering (e.g., resistance to pests and herbicides) will undoubtedly go a long way. Similarly, Sex pheromones can be used for monitoring, mass trapping, and mating disruption. This will serve as a decision support tool, enabling informed decisions. Furthermore, advanced research is necessary to comprehend the mobility and complexity of pesticides in diverse ecosystems.

Pesticides should undergo proper treatment to convert them to non-toxic compounds before being released into the environment. They are often persistent and do not degrade so easily in natural conditions. There must be advanced approaches to reduce pesticide pollution, and scientists are also working in that direction. Environmentally sound management strategies, such as microbial degradation, phytoremediation, and myco-remediation, are a few of the benign methods of bioremediation that are cost-effective and sustainable. However, many external factors influence the metabolic pathways of the microbes. Hence, further studies are needed to establish complete reliance on these methods. Microbial degradation is considered to be a slower process and may not occur efficiently in all cases. Scientists need to search for more potent microbes and novel genes, as traditional microbial degradation is not always successful.

Similarly, it is becoming increasingly important to conduct thorough research to identify suitable enzymes capable of degrading persistent synthetic pesticides. Genetically engineered microbes and biotechnology play an important role in this aspect. The use of pesticide-degrading microbes can certainly be a beneficial way to deal with pollutants in an eco-friendly manner. Hence, it is of utmost importance to further screen the effective strains of various microbes and enzymes to mitigate pesticide risks in the environment and human health. The toxicological background of various insecticides and pesticides is necessary to estimate the effect and pathway of pesticides in that environment.

The potential tradeoffs between conserving biodiversity and protecting against pest resistance must be studied, and this could be one of the noteworthy areas of ecological research in the years to come. Since there have been long-term deleterious effects of insecticide use with their time-to-time applications, the spraying patterns, as well as their frequencies, could be modified to ensure the recovery of soil flora and fauna population to meet the objectives of sustainable crop production.

References

Abhilash P C, Singh N. 2009. "Pesticide use and application: an Indian scenario." *Journal of Hazardeous Material* 165(1–3):1–12. https://doi.org/10.1016/j.jhazmat.2008.10.061.

Abubakar Y, Tijjani H, Egbuna C, Adetunji C O, Kala S, Kryeziu T L, Ifemeje J C and Patrick-Iwauanyanwu K C .2020. "Pesticides, history, and classification," in Natural Remedies for Pest, Disease and Weed Control, edited by Chukwuebuka Egbuna, Barbara Sawicka. *Amsterdam: Academic Press, 29–42.* https://doi.org/10.1016/B978-0-12-819304-4.00003-8.

Agarwal A, Prajapati R, Singh OP, Raza SK, Thakur LK. 2015. "Pesticide residue in water- a challenging task in India." *Environmental monitoring and assessment* 187(54):1- 21. https://doi.org/10.1007/s10661-015-4287-y.

Ahemad M, Khan M S. 2013. "Pesticides as antagonists of rhizobia and the Legume-Rhizobium Symbiosis: A paradigmatic and mechanistic outlook." *Biochemistry and Molecular Biology* 1(4):63-75.

Badji C A., Guedes R N C., Silva A A., Correa A S., Queiroz M E L R and Michereff F M. 2007. "Nontarget impact of deltamethrin on soil arthropods of maize fields under conventional and no-tillage cultivation." *Journal of Applied Entomology* 131(1):50-58.

Basley K, Goulson D. 2018. "Neonicotinoids thiamethoxam and clothianidin adversely affect the colonisation of invertebrate populations in aquatic microcosms." *Environmental Science and Pollution Research* 25(10), 9593–9599.

Beulke S, Brown C D, Fryer C J and Van-Beinum W. 2004. "Influence of kinetic sorption and diffusion on pesticide movement through aggregated soils." *Chemosphere* 57(6), 481–490. https://doi.org/10.1016/j.chemosphere.2004.06.026.

Bhatt P, Zhou X, Huang Y, Zhang W and Chen S. 2021. "Characterization of the role of esterases in the biodegradation of organophosphate, carbamate, and pyrethroid pesticides." *Journal of Hazardous Materials* 411:125026. 10.1016/j.jhazmat.2020.125026.

Brown A W A. 1978. Ecology of Pesticides. John Wiley & Sons, Incorporated, John. 0-471-10790-5, New York.

Cavicchioli R., Ripple W J., Timmis K N., Azam F., Bakken L R., Baylis M., Behrenfeld M J., Boetius A., Boyd P W., Classen A T., Crowther T W., Danovaro R., Foreman C M., Huisman J., Hutchins D A., Jansson J K., Karl D M., Koskella B., Welch D B M., Martiny J B H., Moran M A., Orphan V J., Reay D S., Remais J V., Rich V I., Singh B K., Stein L Y., Stewart F J., Sullivan M B., Oppen M J H V., Weaver S C., Webb E A., Webster N S. 2019. "Scientists' warning to humanity: microorganisms and climate change." *Natural Reviews Microbiology* 17:569–586. https://doi.org/10.1038/s41579-019-0222-5.

Cordeiro E M G, Correa A S and Guedes R N C. 2014. "Insecticide-mediated shift in ecological dominance between two competing species of grain beetles." *PLoS ONE*, 9, e100990. https://doi.org/10.1371/journal.pone.0100990.

David C. Coleman. 2008. "From peds to paradoxes: Linkages between soil biota and their influences on ecological processes," *Soil Biology and Biochemistry* 40(2): 271-289, ISSN 0038-0717, https://doi.org/10.1016/j.soilbio.2007.08.005.

Fountain M T., Brown V K., Gange A C., Symondson W O C and Murray P J. 2007. "The effects of the insecticide chlorpyrifos on spider and Collembola communities." *Pedobiologia* 51(2):147-158.

Garcia F P, Ascencio S C, Oyarzun J C G, Hernandez A C and Alavarado P V. 2012. "Pesticides: classification, uses and toxicity. Measures of exposure and genotoxic risks." *Journal of Research in Environmental Science and Toxicology* 1(11), 279–293.

Gaw S K, Palmer G, Kim N D, Wilkins A L. 2003. "Preliminary evidence that copper inhibits the degradation of DDT to DDE in pip and stone fruit orchard soils in the Auckland region, New Zealand." *Environmental Pollution* 122(1): 1-5. https://doi.org/10.1016/S0269-7491(02)00417-7.

Gevao B, Semple K T and Jones K C. 2000. "Bound pesticide residues in soils: a review." *Environmental Pollution* 108(1), 3–14. https://doi.org/10.1016/s0269-7491(99) 00197-9.

Hardin M R, Benrey B, Coll M, Lamp W O, Roderick G K, Barbosa P. 1995. "Arthropod pest resurgence: An overview of potential mechanisms." *Crop Protection* 14, 3–18.

Harrison S A .1990. "The Fate of Pesticides in the Environment, Agrochemical Fact Sheet # 8," Penn, USA.

Hayward A C., Fegan N., Fegan M., Stirling G R. 2010. "Stenotrophomonas and Lysobacter: Ubiquitous plant-associated gamma-proteobacteria of developing significance in applied microbiology." *Journal of Application Microbiology* 108, 756–770.

Husain S., Siddiqu T., Salem M., Arshad M. and Khalid A. 2003. "Impact of pesticides on soil microbial diversity, Enzyme and biochemical reactions," *Advances in Agronomy*, 102: 159 – 200.

Icoz I, Stotzky G. 2008. "Fate and effects of insect-resistant Bt crops in soil ecosystems." *Soil Biology and Biochemistry* 40(3): 559-586. https://doi.org/10.1016/j.soilbio.2007.11.002.

Iwasa M, Nakamura T, Fukaki K and Yamashita N. 2005. "Nontarget effects of ivermectin on coprophagous insects in Japan." *Environmental Entomology* 34(6):1485-1492. http://dx.doi.org/10.1603/0046-225X-34.6.1485.

Jansch S., Frampton G K., Rombke J., van den Brink P J., and Scott-Fordsmand J J. 2006. "Effects of pesticides on soil invertebrates in model ecosystem and field studies: A review and comparison with laboratory toxicity data." *Environmental Toxicology and Chemistry* 25(9), 2490–250. https://doi.org/10.1897/05-439R.1.

Jindal R and Sharma R. 2019. "Neurotoxic responses in brain of Catla catla exposed to cypermethrin: a semiquantitative multibiomarker evaluation." *Ecological Indicators* 106:105485. https://doi.org/10.1016/j.ecolind.2019.105485.

Johnson M T and Strinchcombe J R. 2007. "An emerging synthesis between community ecology and evolutionary biology." *Trends in Ecology and Evolution* 22 (5): 250–7.

Khan M J, Zia M S and Qasim M. 2010. "Use of pesticides and their role in environmental pollution." *World Academy of Science, Engineering and Technology* 4(12), 122–128.

Liess M. and von der Ohe P C. 2005. "Analyzing effects of pesticides on invertebrate communities in streams." *Environmental Toxicology and Chemistry* 24(4), 954–965. doi: 10.1897/03-652.1.

Lorenz E S. 2009. "Potential Health Effects of Pesticides." Ag Communications and Marketing. 1-8.

Miller G T. 2004. "Sustaining the Earth, 6th edition." Thompson Learning, Inc. Pacific Grove, California, USA.

Pandey S, Singh D K. 2004. "Total bacterial and fungal population after chlorpyrifos and quinalphos treatments in groundnut (Arachis hypogaea L.) soil." *Chemosphere* 55(2): 197-205.

Penca C and Hodges A C. 2017. "Pyriproxyfen treatment terminates *Halyomorpha halys* reproductive diapause, with an indirect mortality effect on its egg parasitoid *Trissolcus japonicus*." *Jounal of Pest Science* 90, 1269–1276.

Prischmann D A, James D G, Wright L C, Teneyck R and Snyder W E. 2005. "Effects of chlorpyrifos and sulfur on spider mites (Acari: Tetranychidae) and their natural enemies." *Biological Control* 33, 324–334.

Relyea R A, Hoverman J T. 2008. "Interactive effects of predators and a pesticide on aquatic communities." *Oikos* 117(11), 1647–1658.

Resende-Silva G A, Turchen L M, Guedes R N C, Cutler G C. 2019. "Imidacloprid soil drenches affect weight and functional response of spined soldier bug (Hemiptera: Pentatomidae)." *Journal of Economic Entomology* 112, 558–564.

Rodrigues A C M, Bordalo M D, Golovko O, Koba O, Barata C, Soares A M V M, Pestana J L T. 2018. "Combined effects of insecticide exposure and predation risk on freshwater detritivores." *Ecotoxicology* 27, 794 802.

Sacramento C A. 2008. Department of Pesticide Regulation "What are the Potential Health Effects of Pesticides?" Community Guide to Recognizing and Reporting Pesticide Problems, 27-29.

Sanchez-Bayo F and Hyne R V. 2011. "Comparison of environmental risks of pesticides between tropical and nontropical regions." *Integrated Environmental Assessment and Management* 7(4):577- 586. https://doi.org/10.1002/ieam.189.

Sharma A., Kumar V., Shahzad B., Tanveer M., Sidhu G P S., Handa N., Kohli S K., Yadav Poonam., Bali A S., Parihar R D., Dar O I., Singh K., Jasrotia S., Bakshi P., Ramakrishnan M., Kumar S., Bhardwaj R and Thukral A K. 2019. "Worldwide pesticide usage and its impacts on ecosystem." *SN Applied Sciences* 1(1446):1–16. https://doi.org/10.1007/s42452-019-1485-1.

Sigler W V, Turco R F. 2002. "The impact of chlorothalonil application on soil bacterial and fungal populations as assessed by denaturing gradient gel electrophoresis." *Applied Soil Ecology* 21(2): 107-118.

Silva V., Mol H G., Zomer P., Tienstra M., Ritsema C J., and Geissen V. 2019. "Pesticide residues in European agricultural soils-A hidden reality unfolded." *Science of the Total Environment* 653:1532–1545. https://doi.org/10.1016/j.scitotenv.2018.10.441.

Subhani A., El-ghamry A., Changyong H and Jianming Xu. 2000. "Effects of Pesticides (Herbicides) on Soil Microbial Biomass - A Review." *Pakistan Journal of Biological Sciences* 3: 705-709.

Tudi M, Daniel Ruan H, Wang L, Lyu J, Sadler R, Connell D, Chu C and Phung D T. 2021. "Agriculture development, pesticide application and its impact on the environment." *International Journal of Environmental Research and Public Health* 18(2):1112. https://doi.org/10.3390%2Fijerph18031112

Wang Y, Lai A, Latino D, Fenner K, Helbling D E. 2018. "Evaluating the environmental parameters that determine aerobic biodegradation half-lives of pesticides in soil with a multivariable approach." *Chemosphere*. 209:430-438. https://doi.org/10.1016/j.chemosphere.2018.06.077.

Williams T., Valle J. and Vinuela E. 2003. "Is the naturally derived insecticide Spinosad® compatible with insect natural enemies?." *Biocontrol science and Technology* 13(5):459- 475.

Woodrow J E, Gibson K A and Seiber J N. 2018. "Pesticides and related toxicants in the atmosphere." *Reviews of Environmental Contamination and Toxicology* 247, 147–196. https://doi.org/10.1007/398_2018_19.

World Health Organization. 2009. Children's Health and the Environment. WHO Training Package for the Health Sector-World Health Organization. Geneva: World Health Organization.

Wu Y, Li W, Yuan M and Liu X. 2020. "The synthetic pyrethroid deltamethrin impairs zebrafish (Danio rerio) swim bladder development." *Science of Total Environment* 701:134870. https://doi.org/10.1016/j.scitotenv.2019.134870.

Yadav I C and Devi N L. 2017. "Pesticides classification and its impact on human and environment." *Environmental Science and Engineering* 6, 140–158.

Zwieten L V, Rust J, Kingston T, Merrington G, Morris S. 2004. "Influence of copper fungicide residues on occurrence of earthworms in avocado orchard soils." *Science of the total Environment* 2004; 329(1-3): 29-41.

Chapter 8

Regulatory and Policy Issues in Insecticide Use

Vikram[1]
Geetika[1]
Kapil Attri[2]
Bhawani Singh Meena[3]
and Ashish Sheera[3,*]
[1]Department of Agriculture, Mata Gujri College, Fatehgarh Sahib, Punjab, India
[2]Division of Sericulture, SKUAST-J, Main Campus, Chatha, Jammu (J&K), India
[3]College of Agriculture, Baseri, Dholpur (SKN Agriculture University, Jobner), Rajasthan, India

Abstract

Insecticide use is an important tool in agricultural and public health efforts to control insect pests. However, these chemicals have negative impacts on human health and the environment. As a result, regulatory and policy issues related to insecticide use have become increasingly important in recent years. We are exploring the issues related to the development and implementation of regulations governing the use of insecticides, the evaluation of hazards coupled with their consumption, and the development of alternative pest control strategies. It will also consider the role of stakeholders, including government agencies, industry groups, and environmental organizations, in shaping policy related to insecticide use. Finally, the chapter will examine some of the challenges associated with regulating and managing insecticide use, including the development of resistance to insecticides among target pests. The need for ongoing monitoring and evaluation, as well as the

* Corresponding Author's Email: sheeraashish@gmail.com

In: Insecticides
Editors: Archana Anokhe and Govindaraj Kamalam Dinesh
ISBN: 979-8-90134-022-6

potential for unintended consequences associated with alternative pest control strategies. This compilation presents a summary of significant regulatory and policy issues related to insecticide use, providing insights into how these issues might be addressed in the future.

Keywords: insecticide, human health, environment, key regulatory and policy issues

Introduction

Most agricultural production industries consume pesticides to prevent or reduce pests; the quality and quantity of produce can be improved by reducing losses, even in terms of cosmetic attraction, which is often of importance to users (Oerke and Dehne 2004; Cooper and Dobson 2007). In addition to improving nutritional value, pesticides can make food healthier and safer (Boxall, 2001; Narayanasamy, 2006). Furthermore, pesticides may also provide several benefits to the general public; however, the public often overlooks these advantages (Damalas 2009). This means that pesticides can be considered cost-effective, labour-saving, and effective pest management tools commonly used in agriculture (Damalas and Eleftherohorinos 2011). Globally, more than a thousand pesticides are used to minimize crop losses due to biotic stresses, including chemical and biological ones.

Pest infestations have a significant impact on agricultural production in developing countries. Approximately $ 42.66 million is lost in India each year due to pests (Sushil, 2016). Several concerns have been raised regarding the impact of chemical pesticides on soil and the environment, as well as the presence of residues in food. Several other issues concern the structure and regulations of the pesticide industry. Several of these issues are discussed briefly here in the context of the Indian pesticide industry (Subash et al., 2017). Numerous reports indicate that pesticides have a negative impact on non-target organisms, including bees, birds, and aquatic species (Beketov et al., 2013), contributing to the global biodiversity crisis. The difficulties faced by farmers, especially small-sized ones, when it comes to complying with stricter legislative frameworks are remarkable despite the benefits that can be gained. When it comes to pesticides, the risk of worker exposure is extremely complex compared to the risks associated with other aspects of occupational health and safety (OHS). There is a need to fill the gap, as mentioned above, by developing an easy-to-use tool to assess occupational risks associated with pesticide use in agricultural activities. The Agricultural Health Study (AHS) served as the starting point for developing the procedure, followed by the creation of a preliminary risk assessment tool based on the AHS approach (Fargnoli et al., 2019). This tool included a simplified model that farmers could apply in their practice.

Regulations for Pesticide Use

Adopt policies and legislation to regulate pesticides and ensure their practical use, marketing, and regulation throughout their lifecycle. Establishing services in the areas of education, advisory, extension, and healthcare, based on FAO and WHO guidelines, as well as legally binding instruments when applicable, is essential for the development of such services. Local governments must implement pesticide application and personal protective equipment programs in accordance with local needs, socioeconomic factors, literacy levels, climatic conditions, and the availability and affordability of pesticides. To determine pesticide equivalence, refer to the Manual on FAO and WHO Specifications for Pesticides. As much as possible, incorporate international standards and guidelines into national and regional legislation, promoting a harmonized approach (region- or country-wise) in pesticide registration, procedures, and assessment standards. Analyze the amount, quality, quantity, and use of pesticides imported, exported, manufactured, and used to determine their potential effects on human and animal health, as well as the environment.

Additionally, monitor pesticide use trends for economic and other purposes. Pesticide residues in food should be regulated and monitored in accordance with Codex Alimentarius recommendations. If Codex standards are not available, use national or regional standards that are equivalent or comparable to Codex standards. Technical trade barriers must be prevented in a manner consistent with WTO requirements.

India enacted pesticide regulations in 1968. The law has been amended and updated since 2008, in accordance with current pesticide usage scenarios. Reforms, however, are proving difficult for the government. According to the Insecticides Act, 1968, and the Insecticides Rules, 1971, the use of insecticides is regulated. However, this single law lacks sufficient provisions to prevent farms from being contaminated or to minimize risks to farm workers (Reddy 2022). Among Indian farmers, public health workers, industries, and construction companies, pesticides are widely used. For pesticides to be effective and safe, they must be managed and regulated throughout their entire life cycle, encompassing manufacturing, importing, packaging, labeling, pricing, storage, advertising, sales, transportation, distribution, and use. It is vital that safe and effective pesticides are available and that the risks to humans, animals, and other living organisms are minimized. According to developments in toxicology (Chari, 2018), there are currently no mechanisms for periodic inspection of pesticides.

Human Health Protection Law

According to the Indian regulatory framework, pesticides can be registered, imported, and sold after they have been registered and licenses have been obtained. Such an application can be granted or rejected by the Central Insecticides Board. There is no precise mechanism to challenge the registrations granted under the Act, despite its provision for a registration procedure. State governments are limited in their ability to regulate pesticides. Licenses for manufacturers, sellers, distributors, stockists, and exhibitors of pesticides may be issued or revoked for a limited time. If a safety concern arises, states can ban pesticides for up to 60 days, with some extending their bans for an additional 30 days. Even if their use poses an imminent danger to farmers, certain pesticides remain available for sale for up to 60 days. Additionally, implementing a pesticide ban for such a short period poses challenges (Bonvoisin et al., 2020).

Grievance Mechanisms

Due to their hazardous nature and necessity, pesticides are unique products. To prevent misuse, abuse, and accidental exposure to pesticides, they must be adequately regulated and monitored. In India, there is no grievance redressal mechanism for pesticide regulation. The exposure to pesticides and the risk to their lives are inevitable for pesticide users. However, pesticide laws and policies remain silent on the mechanisms that provide users with access to training and information in the event of an accident. There is no legal grievance mechanism available to third parties who may be exposed to pesticides as a result of their employment or any other circumstance in their lives. Several pesticides can pose risks to humans, wildlife, and other organisms, including neurological and reproductive health issues, hormonal disruptions, and cancer (Nicolopoulou-Stamati et al., 2016).

Collection of Data and Information

The Collection of Statistical Act (2008) and pesticide regulations do not mandate the collection of information and data on pesticide usage and its impact on people, the environment, and ecology. The National Crime Records Bureau maintains a database of pesticide poisoning cases under the Code of Criminal Procedure. The NCRB has limited responsibilities due to its limited mandate (Reddy, 2016), and it faces several challenges.

Insecticide Registration and Quality Control

Pesticide import, registration, production, sale, transfer, distribution, and use are regulated by the Insecticides Act (1968) and Insecticides Rules (1971). Before pesticides can be sold or manufactured, they must be registered with the Central Insecticides Board and the Registration Committee (CIB&RC). To obtain approval for manufacturing or importing, applicants must submit data on various aspects, including chemical composition, toxicity, and bio-efficacy. The authentic report of research and development organizations is also used as a valid source of information on some topics (such as the bio-efficacy of pesticides). After verifying the application's validity, the Committee issues a registration number and certificate. As of June 2017, CIB&RC lists 279 products (265 chemicals and 14 bio-pesticides) and 658 formulations. Testing pesticides at different stages is regulated, and procedures are in place. The State Pesticide Testing Laboratories (SPTLs) test samples for quality control during manufacturing and at the point of sale, while the Central Insecticide Laboratory (CIL) tests samples submitted by government agencies and officers. Approximately 2.5 to 3 percent of samples tested between 2008-09 and 2012-13 were mislabeled. The country has banned 28 pesticides, withdrawn eight pesticides, and restricted the use of 13 others. Twelve pesticides will be banned from January 2018 and six more from December 2020, according to a committee of experts (DAC&FW, 2016).

Environmental Issues

As a result of stringent environmental regulations across the globe, domestic manufacturers do not put enough effort into R&D due to high costs, innovation and product diversification are important, farmers are not aware of how to use pesticides safely, new products require long gestation periods, and product quality assurance is difficult to achieve (FICCI, 2015) besides protecting farmers from substandard products, pesticide safety programs should be implemented, along with measures to mitigate potential health and environmental impacts. It is impossible to rule out the possibility that substandard products may be present; therefore, there should be an emphasis on enforcing quality inspections at the point of sale and protecting farmers through consumer forums. Crop losses can be controlled with bio-pesticides, and environmental damage can be reduced. An estimated 3% of the country's pesticide market is devoted to bio-pesticides. Under the Insecticide Act 1968, 14 bio-pesticides have been registered in India. Since 1996-97, the capacity of the biopesticide industry has grown from 219 tonnes to 683 tonnes, and then to around 3000 tonnes in 2015-16 (Sinha and Biswas, 2008; DAC&FW, 2017). (Birthal 2003) found that 66 percent of cotton pesticide use can be reduced through the use of bio-pesticides when integrated pest management is employed.

Consequently, biopesticides can provide reliable, sustainable, and environmentally friendly alternatives to chemical pesticides. Despite this, the market for bio-pesticides has not developed as rapidly as it could. All levels of the supply chain should develop skills and facilities for storing bio-pesticides (Subash et al., 2017), and have also suggested providing fiscal incentives for producing and using bio-control agents.

Benefits of Insecticide Use

The benefits derived from the use of pesticides are primarily derived from their direct effects. The primary advantages of killing caterpillars that feed on cabbage are higher yields and improved quality of the cabbage. Among the 26 primary benefits resulting from these three main effects are the protection of recreational turf and the saving of human lives. As a result of the primary benefits, secondary benefits emerge that are less apparent or less immediate. It may be subtle, less intuitively obvious, or long-term in nature. In the context of secondary benefits, establishing cause and effect has proven more difficult, but this does not mean that they cannot be powerful justifications for pesticide use, as farmers have made significant gains in food production through their use (Aktar et al., 2009). It has primarily been done to reduce agricultural losses due to pest activity, leading to improved yields and greater access to food throughout the year at reasonable prices. Most countries have increased their agricultural productivity by using pesticides. The potential risks of consuming organic crops with very low pesticide residues have long been regarded as outweighing those of diets containing fresh fruits and vegetables (Brown 2004). The quality of life and longevity can both be improved by improved nutrition and reducing drudgery (WHO, 2004). The nutrition, safety, and affordability of a healthy diet can have a significant impact on life expectancy. However, it can also improve medical care, drug treatments, and hygiene. Agricultural losses have been reduced due to pest control efforts, resulting in improved yields and greater access to affordable food throughout the year. A large portion of humankind would be exposed to unacceptably risky conditions without chemical control methods, according to the World Health Organization (Maksymiv 2015).

Hazards of Pesticides

Numerous health and environmental concerns are associated with pesticide use. Because pesticides are vital to human survival, toxic pesticides contaminate food and have serious health consequences. Pesticides: Benefits and Hazards. 98% of insecticides sprayed and 95% of herbicides applied are not intended for their intended targets, such as non-target species, soil, air, or water. Ingestion, inhalation, or dermal absorption of pesticides can cause toxicity. By prolonged exposure to these chemicals, the following diseases can develop: - Neurological, psychological, and behavioural dysfunction; - Hormonal imbalances, leading to infertility and breast pain; - Immune system dysfunction; - Reproductive system defects; - Cancer; - Genotoxic; - Blood disorders. Pesticides can contaminate vegetation, soil, and water. In addition to insects or weeds, other organisms can also be poisoned by pesticides, including birds, fish, and beneficial insects. Pesticides are associated with various health effects, which have been reviewed in recent articles and reports (Kubrak, 2012). Many pesticides to which the general population is chronically exposed may be carcinogenic, neurotoxic, reproductive toxins, and immunotoxic to animals, as demonstrated through extensive toxicological studies (Gonzalez et al., 2005). Cells from Chinese Hamster ovary (CHO) exposed to 2,4-D showed DNA damage. The development of neurodegenerative diseases has been linked to pesticides. Numerous studies have shown that pesticides affect biochemical parameters, especially protein metabolism (Li Z. H., 2011), the endocrine system, and reproductive functions. Water used for public consumption can be contaminated with pesticides, whether from surface or groundwater sources (Cerejeira et al., 2003). The chemical constituents of agricultural production are directly absorbed by aquatic organisms via surface runoff or indirectly by trophic chain interactions (Maksymiv, 2015).

Risk Assessment

Ecosystems Lack Drivers of Risk

According to Knobel (2014), Current regulations have failed to protect biodiversity due to inaccurate predictions of exposure and effects. Models and test systems must consider the following factors in order to improve predictions based on specific, spatially explicit real-life scenarios (Schäfer 2018).

1) Ecological context: The ecological properties of the entire ecosystem, including organism sensitivity and functions, need to be protected.
2) Landscape context: Ecosystems are embedded in certain situations. Additionally, site-specific aspects, including connectivity and stressors, are also considered in conjunction with general environmental conditions.
3) Management context - There is a possibility that mixtures of compounds may occur in soils and water bodies, exacerbated by risk management practices based on farmers' use patterns, including minimum spray distances to field margins and the sequence of pesticide use.

Corrections to the Current System in the Short Term

In the short term, it is necessary to correct the current system (Schäfer 2012) by considering ecological, landscape, and management uncertainties, and by aligning predictions and observations of ecosystem effects. It would be beneficial if the authorization process included a cost-benefit analysis, as well as a comparison of the risks associated with a pesticide product to those of already authorized pesticides. The most environmentally friendly product could be authorized using comparative assessments. The cost-benefit ratio can be analyzed by comparing potential yield losses with environmental quality benefits. The cost-benefit analysis will fail if the product decreases yield losses while deteriorating the environment. However, quantification may prove challenging for new compounds (Schäfer et al., 2019).

Human Health and Pesticides

Human health risks posed by pesticides can be complex and particularly accurate when considering the various periods and levels of exposure (in terms of toxicity), types of pesticides used, mixtures used in the field, geographical characteristics, and meteorological characteristics of agricultural areas where pesticides are applied (Pastor et al., 2003). Differences between the pesticide sprayers, preparers of the mixtures, as well as those living in areas surrounding sprayed fields, storage facilities, greenhouses, or open fields, contribute to these variations. As a result, if pesticide toxicity and exposure both affect human health, moderately toxic pesticides are more likely to pose a serious risk than highly toxic pesticides. There is considerable scientific controversy surrounding pesticide residues in food and water (Magkos et al., 2006; Damalas and Eleftherohorinos, 2011).

Management Strategies

Practices of General Management

Pesticide use phases have been clarified in European regulations (EC 2000, EC 2009a, and EC 2009b). Pesticides can be prevented from entering water from point sources, and diffuse sources (such as drainage, spray drift, and runoff) can be reduced. Stakeholders' awareness of risks is the single most important factor in an effective risk management program. There is a need for practical and understandable tools to establish an accurate diagnosis and to select effective mitigation measures. It is more challenging to prevent pesticide entry into the water through diffuse sources because not all relevant factors can be controlled (for example, weather, the time interval between pesticide application and the first significant rainfall event). A weather prediction will now help us prevent unnecessary delays.

Indicators for Pesticide RM

Based on a limited number of variables, indicators depict pictures (e.g., a class or number). It is possible to assess risk quantitatively or qualitatively with the help of risk indicators. Indicators of pesticide risk can simplify the assessment of health and environmental risks associated with agricultural pesticides. When resources and data availability are limited, indicators may be beneficial (Feola et al., 2011). Risk indicators are designed to help risk managers, who are not necessarily scientists, evaluate risks and make informed decisions. However, many indicators are inaccurate, inappropriate for their contexts, use misleading proxies of risk, lack scientific rigour, or lack adequate validation (Devillers et al., 2005).

Ecological Engineering

Agricultural plots export pesticides into surface water and groundwater. The discharge zone can cause groundwater contamination of surface waters, explaining why streams remain polluted (Milosevic et al., 2012). The only way to reduce pesticide pressure on groundwater is to reduce pesticide use. Surface water mitigation strategies include, however, some complementary measures (Gregoire et al., 2009).

The biodegradation and retention of pesticides occur when pesticides are dissipated in buffer zones such as constructed wetlands and buffer strips, where complex interactions occur between abiotic (physicochemical properties of pesticides, residence times, and water flow, adsorption and desorption kinetics) and biotic factors (plant cover, abundance, microbial diversity, and activity). Since buffer zones contain low levels of pesticides, one limitation of their purifying capability is their low concentration of pesticides. The application of ecological engineering approaches that improve the interception of pesticide fluxes and the degradation capacity of environmental microbial communities (soil and sediment) can (1) Buffer zones should be made more pesticide-resistant, (2) prevent pesticides from persisting and dispersing in adjacent environments by promoting their biodegradation in buffer zones (Pesce et al., 2009).

Labelling Requirements

What Is a Label

Labels are written, printed, or graphic materials attached to product containers. Government regulatory agencies and industry have worked together for years on this research and evaluation. Labels are mandatory on product packages. Technical leaflets or brochures accompanying a container may provide additional information, which should be referred to on the label if available.

Labels Must Be Clear and Accurate

Labels are the main point of contact between manufacturers/suppliers and end users. In addition to conveying essential safety information and use recommendations, labels are legal documents. A simple and direct message on the label can be the key to getting the user to "read the label." National laws primarily regulate the content level. Besides these, individual companies have their standards, as well as international standards, such as the FAO International Code of Conduct for the Distribution and Use of Pesticides. All of these regulations and standards should be followed on labels.

Label Content

The label of an item provides detailed information about the product and its safe and efficient use. National laws primarily regulate the content level. Occasionally, there may be considerations related to multinational companies or specific regions.

a) Identification of the product category, such as an herbicide, insecticide, or fungicide.
b) Refer to the International Formulation Coding System for the type of formulation and its corresponding code.
c) Active ingredient, name (ISO) or other locally used common name, or in the absence of either, the chemical name as used by ISO or IUPAC, and content. This should generally be expressed as "contains x g ai per kg" (for solids, viscous liquids, aerosols, or volatile liquids), "contains x g ai per litre" (for other liquids), or just "y%."
d) Net contents of the pack. Metric units should be used if the country does not or only partly use them (e.g., liters, grams, kilograms). It is essential to use local units whenever possible, but metric units should also be employed in such situations.

Information about safety

- Before opening the package, read the safety instructions.
- Warning symbols for handling, transporting, and storing.
- Risk category/symbol. Depending on the product's toxicity, it may need to be classified.

Use of colour

- As a general rule, red is used for highlighting warning phrases, hazard symbols, and safety precaution instructions.
- Labels should be primarily black on white backgrounds to provide the best contrast and ease of reading.
- Generally, colour enhances the attractiveness of leaflets and brochures. Understanding will increase when things are shown in their true colours. Contrasting colours can highlight important parts of drawings. The message can be distracted by too many or too intense colours.
- There is just as much importance to colour contrast as there is to colour itself. It is always best to display red on a white background and never on another colour.
- To maintain contrast, use intense colours, such as black against white, black against yellow, red against white, green against white, and white against blue, on neutral backgrounds.

- Using the WHO classification of formulation hazards, the FAO introduced a colour coding scheme.
- The FAO uses colour codes to indicate toxicity levels. Labels must include the relevant colour band as specified by local regulations. Labels cannot use the same colour elsewhere, according to the FAO (1995).

International Agreements

Minamata Convention on Mercury

The Minamata Convention on Mercury protects human health and the environment from the adverse effects of mercury. A decision was reached during the fifth and final session of the Intergovernmental Negotiating Committee in Geneva on 19 January 2013. The Convention highlights several key points, including the prohibition on new mercury mines, the phase-out of existing mercury mines, and the international regulation of artisanal and small-scale gold mining.

Information, Public Participation and Access to Justice in Environmental Matters - Aarhus Convention

- On 25 June 1998, the Fourth Ministerial Conference of Environment for Europe in Aarhus, Denmark, adopted this UNECE Convention.
- To future generations, we have an obligation regarding environmental and human rights.
- The participation of all stakeholders is essential to achieving sustainable development.
- Protects the environment and ensures government accountability.
- A study of public-private interactions in a democratic environment.
- The Convention addresses a fundamental issue in the relationship between people and governments. There is more to convention than just the environment. It is also about government accountability, transparency, and responsiveness.
- As part of the Aarhus Convention, public participation, access to information, and access to justice are guaranteed. The Aarhus Convention also introduces a new way for the public to participate in international negotiations.

Persistent Organic Pollutants Convention of Stockholm

Globally, the Stockholm Convention on Persistent Organic Pollutants protects human health and the environment from persistent organic pollutants that remain intact for extended periods, are widely distributed geographically, and accumulate in fatty tissues over time. There is evidence that persistent organic pollutants (POPs) may cause cancer, congenital disabilities, immune system malfunctions, and even intelligence decline in some cases. Governments cannot protect their citizens or their environment alone from POPs, which travel over long distances. The Stockholm Convention, adopted in 2001, requires Parties to eliminate or reduce the release of POPs. A UN Environment Programme office is located in Geneva, Switzerland, where the program is administered.

Convention on Prior Informed Consent for the Trade of Pesticides and Hazardous Chemicals

In Rotterdam, the Netherlands, a Conference of Plenipotentiaries adopted the Convention on 10 September 1998. On 24 February 2004, the Convention entered into force.

The Convention has the following objectives:

- The international trade of certain hazardous chemicals should be governed by a shared responsibility and cooperation between parties to protect human health and the environment.
- Guide how to import and export these hazardous chemicals, facilitate information exchange about them, and communicate their importation and exportation to the public.
- Under the Convention, the Prior Informed Consent (PIC) procedure must be implemented in a manner that is legally compliant. UNEP and FAO initiated the voluntary PIC procedure in 1989, which ended on 24 February 2006.

References

Aktar MW, Sengupta D, Chowdhury A. Impact of pesticides use in agriculture: their benefits and hazards. *Interdisciplinary Toxicology* (2009) 2(1): 1–12.

Beketov MA, Keford BJ, Schäfer RB, Liess M. Pesticides reduce regional biodiversity of stream invertebrates. *Proc Natl Acad Sci* (2013) USA110:11039–11043.

Birthal PS. Economic Potential of Biological Substitutes for Agrochemicals. Policy Paper 18, ICAR-National Institute of Agricultural Economics and Policy Research, New Delhi (2003).

Bonvoisin T, Leah U, Duleeka K. Suicide by Pesticide Poisoning in India: a review of pesticide regulations and their impact on suicide trends. BMC Public Health 20 (2020),

Boxall RA. Post-harvest losses to insects—a world overview. *Int Biodeter Biodegr* (2001) 48:137-152.

BrownI. UK Pesticides Residue Committee report. (2004). Available at: http://www. pesticides.gov.uk/uploadedfiles/Web_Assets/PRC/PRCannualreport2004.pdf.

Cerejeira MJ, Viana P, Batista S, Pereira T, Silva E, Valerio MJ. Pesticides in Portuguese surface and ground waters. *Water Res* (2003) 37 (5): 1055-1063.

Chari M. Eight terrifying things we found about pesticide regulation and use in India (2018). Scroll.in,https://scroll.in/article/877049/eight-terrifying-things-we-found-about-pesticideregulation-and-use-in-india.

Cooper J, Dobson H. The benefits of pesticides to mankind and the environment. *Crop Prot* (2007) 26:1337-1348.

DAC & FW. Banning of Pesticides Order. Department of Agriculture Cooperation, Ministry of Agriculture and Farmers Welfare, New Delhi (2016).

DAC & FW. Pesticide Wise Consumption of Indigenous Pesticides during 2010-11 to 2016-17. Directorate of Plant Protection, Quarantine & Storage, Department of Agriculture Cooperation, New Delhi (2017).

Damalas CA. Understanding benefits and risks of pesticide use. *Sci Res Essays* (2009) 4: 945-949.

Damalas CA, Eleftherohorinos IG. Pesticide Exposure, Safety Issues, and Risk Assessment Indicators. *Int J Environ Res* (2011) 8:1402-1419; doi:10.3390/ijerph8051402.

Devillers J, Farret R, Girardin P, Riviere JL, Soulas G. Indicateurs pour evaluer les risquesliesallutilisation des pesticides. Tec & Doc, Lavoisier (in French), Paris, France (2005).

Fargnoli M, Lombardi M, Puri D, Casorri L, Masciarelli E, Colosio C. The Safe Use of Pesticides: A Risk Assessment Procedure for the Enhancement of Occupational Health and Safety (OHS) Management. *Inter J of Env Res and Public Heal* (2019) 16: 310-320.

Feola G, Rahn E, Binder CR Suitability of pesticide risk indicators for less developed countries: a comparison. *Agric Ecosyst Environ* (2011).142:238–245.

FICCI Ushering in the 2nd Green Revolution: Role of Crop Protection Chemicals. Federation of Indian Chambers of Commerce and Industry, New Delhi (2015).

Global strategy on diet, physical activity and health. Fifty-seventh World Health Assembly. WHO, Geneva, (2004).

Gonzlez M, Soloneski S, Reigosa MA, Larramendy ML. Genotoxicity of the herbicide 2,4-dichlorophenoxyacetic and a commercial formulation, 2,4-dichlorophenoxyacetic acid dimethylamine salt. I. Evaluation of DNA damage and cytogenetic endpoints in Chinese Hamster ovary (CHO) cells. *Toxicol In Vitro* (2005) 19: 289-97.

Gregoire C, Elsaesser D, Huguenot D, Lange J, Lebeau T, Merli A, Mose R, Passeport E, Payraudeau S, Schutz T, Schulz R, Tapia-Padilla G, Tournebize J, Trevisan M, Wanko A. Mitigation of agricultural nonpoint-source pesticide pollution in artificial wetland ecosystems. *Environ Chem Lett* (2009) 7(3):205–231.

Knabel A, Meyer K, Rapp J, Schulz R. Fungicide Feld concentrations exceed focus surface water predictions: urgent need of model improvement. *Environ Sci Technol* (2014) 48:455–463. https://doi.org/10.1021/es404 8329.

Kubrak OI, Atamaniuk TM, Husak VV, Drohomyretska IZ, Storey JM, Storey KB, Lushchak VI Oxidative stress responses in blood and gills of Carassiusauratus exposed to the mancozeb containing carbamate fungicide Tattoo. *Ecotoxicol Environ Saf* (2012). 85: 37-43.

Li ZH, Velisek J, Zlabek V, Grabic R, Machova J, Kolarova J, Li P. Chronic toxicity of verapamil on juvenile rainbow trout (*Oncorhynchus mykiss*): Effects on morphological indices, hematological parameters and antioxidant responses. *J Hazard Mater* (2011). 185:870-880.

Magkos F, Arvaniti F, Zampelas A (2006) Organic Food: Buying more safety or just peace of mind? A critical review of the literature. Crit. Rev. Food Sci. Nutr. 46:23-55.

Maksmyliv I. Pesticides: Benefits and Hazards. Journal of Vasyl Stefanyk Precarpathian National University (2015) 2:70-76.

Milosevic N, Thomsen NI, Juhler RK, Albrechtsen HJ, Bjerg PL. Identification of discharge zones and quantification of contaminant mass discharges into a local stream from a landfill in a heterogeneous geologic setting. *J Hydrol* (2012).446– 447:13–23.

Narayanasamy P. Postharvest Pathogens and Disease Management; John Wiley& Sons: New York, NY, USA (2006).

Nicolopoulou-Stamati PS, Maipas C, Kotampasi P, Stamatis Hens, L. Chemical Pesticides and Human Health: The Urgent Need for a New Concept in Agriculture. *Frontiers in Public Health,* https://www.frontiersin.org/articles/10.3389/fpubh.2016.00148/full.

Oerke EC, Dehne HW Safeguarding production-losses in major crops and the role of crop protection. *Crop Prot* (2004) 23:275-285.

Pastor S, Creus A, Parrón T, Cebulska-Wasilewska A, Siffel C, Piperakis S, Marcos R Biomonitoring of four European populations occupationally exposed to pesticides: Use of micronuclei as biomarkers. *Mutagenesis* (2003). 18:249-258.

Pesce S Martin-Laurent F Rouard N, Montuelle B. Potential for microbial diuronmineralisation in a small wine-growing watershed: from treated plots to lotic receiver hydrosystem. *Pest Manag Sci* (2009) 65(6):651–657.

Reddy Donthi N, Dileep Kumar A. The New Pesticide Bill Protects Industry but Leaves Farmers at Risk Caravan Magazine food (2022).

Reddy, Dorothi N. Challenges of Data and Management in Public Health,' Economic and Political Weekly. (2016) 12; 45-46.

Schäfer A, Filser J, Frische T. The silent spring—on the need for sustainable plant protection. German National Academy of Sciences Leopoldina, Halle (2018).

Schäfer RB, Von der Ohe P, Rasmussen J. Thresholds for the effects of pesticides on invertebrate communities and leaf breakdown in stream ecosystems. *Environ Sci Techno* (2012) l46:5134–5142.

Schäfer RB Liess M, Altenburger RJ, Filser U, Hollert H, Nickoll J, Schäfer A, Scheringer A Future pesticide risk assessment: narrowing the gap between intention and reality. *Environ Sci Europe* (2019) 31:21.

Sinha B, Biswas L. Potential of Biopesticides in Indian Agriculture vis-a-vis Rural Development. India Science and Technology: NISTADS, New Delhi (2008).

Subash SP, Prem Chand, Pavithra S, Balaji SJ, Suresh Pal. Pesticide Use in Indian Agriculture: Trends, Market Structure and Policy Issues. ICAR – National Institute of Agricultural Economics and Policy Research (2017).

Sushil, SN. Emerging Issues of Plant Protection in India. Natural Resource Management: Ecological Perspectives. *International Conference*, SKUAST, Jammu (2016).

Chapter 9

Insecticides: Ecological Consequences and Health Hazards

Roopa Rani Samal[1,*]
Kungreiliu Panmei[2]
P. Lanbiliu[2]
Abhay Pratap Singh[2]
Manu Sankar[2]
and Sarita Kumar[2,†]

[1]Department of Zoology, Dyal Singh College, University of Delhi, New Delhi, India
[2]Department of Zoology, Acharya Narendra Dev College, University of Delhi, New Delhi, India

Abstract

Insecticides have become a necessary component of contemporary life, protecting domestic settings, agricultural land, food storage facilities, floral diversity, and human health from deleterious insect pests. These not only cause extensive harm to various food and ornamental crops but also spread infectious diseases that are hazardous to humans. A variety of insecticides with diverse chemical ingredients and in several formulations have been formulated to manage these pests. Despite the extensive global usage of these, worth $38 billion annually, novel insecticide formulations are continually being developed due to their negative impact on the environment, association with various acute and chronic human ailments, and the rise in resistant insect populations.

* Corresponding Author's Email: rupasamal29@gmail.com
† Corresponding Author's Email: sarita.sanjay90@gmail.com; saritakumar@andc.du.ac.in

In: Insecticides
Editors: Archana Anokhe and Govindaraj Kamalam Dinesh
ISBN: 979-8-90134-022-6

Furthermore, the continual application of persistent and non-biodegradable insecticides has contaminated numerous elements in the atmosphere, including those in water, air, and soil ecosystems. These adverse effects have led to a drastic decline in beneficial creatures, such as predators, pollinators, and earthworms, as well as a shift in the richness of soil microbes. Additionally, the persistent nature of these insecticides has allowed them to infiltrate several food chains and bioaccumulate at higher trophic levels, including those of humans and other large mammals. The consequences, as mentioned earlier, of indiscriminate insecticide use illustrate the adverse effects of this practice on several ecosystem elements. Thus, insecticides need to be used in a manner that safeguards the environment and mitigates any potential health risks to humans and other organisms associated with them.

The current chapter focuses on the impact of insecticides on the ecosystem, humans, and non-target as well as beneficial organisms, such as earthworms, fish, amphibians, birds, predators, and pollinators. In addition, the Integrated Pest Management (IPM) approach, which employs a harmonious blend of diverse control measures, including cultural methods, use of resistant genotypes, physical measures, mechanical agents, and balanced use of insecticides, has been discussed in depth to combat the harmful effects of insecticides by reducing the quantity and frequency of insecticide applications. Cutting-edge techniques, such as biotechnology and nanotechnology, which may aid in the development of safer insecticides, have been highlighted. The necessity of community awareness and other extension initiatives that may acquaint and inspire farmers to adopt frontline IPM tactics is recommended.

Keywords: insecticides, environmental impact, health hazards, resistance, IPM

Introduction

Pesticides are chemicals, individual or mixtures, that are used to avert, destroy, deter, or mitigate pests, including insects, weeds, and plant pathogens (USEPA, 2022). They have a long history of use in agriculture, contributing to increased crop production and effective pest control. Since prehistoric times, various chemicals in diverse forms have been utilized for pest control. Ancient civilizations, such as the Sumerians, employed sulfur-based compounds to control insects and mites over 4,500 years ago. Pyrethrum, derived from *Chrysanthemum cinerariaefolium* flowers, has been employed as an insecticide for more than 2,000 years. Various inorganic and organic substances were widely used for pest control until the 1940s (Unsworth, 2010).

The development of pesticides, however, accelerated during World War II (1939-1945) to boost food production and explore them as budding agents of chemical warfare (Gupta, 2007). This period saw the rise of synthetic pesticides, especially organochlorines and organophosphates (DDT, lindane, endrin, dieldrin, endosulfan, aldrin, parathion, and 2,4-D). In the 1950s, the use of pesticides in agriculture was widely perceived as highly beneficial, without raising significant concerns among cultivators regarding the risks to the environment, non-target organisms, crops, and human health (Unsworth, 2010).

Concerns about the indiscriminate use of pesticides began to emerge in the 1960s. Rachel Carson's book "Silent Spring" (1962) highlighted the problems associated with the use of pesticides. It raised awareness about their impact on human health and the environment. Subsequent studies by Ratcliffe further demonstrated the adverse effects of DDT on raptor nests, leading to a decline in eggshell thickness (Ratcliffe, 1967). These factors, combined with the emergence of pest resistance and accumulating evidence on the effects of pesticides, culminated in the ban on DDT in the US in 1972 and subsequent discontinuation in other countries (Levengood et al., 2007).

In the following decades, concerns about the environmental and health impacts of pesticides led to the development of more selective pesticides and research focused on finding safer chemicals, as well as the integration of pest management (IPM) systems. IPM used measures that involved natural predators/parasites to control pests. Along with the pesticide applications, which were timed to coincide with the vulnerable periods in the pest life cycle, this reduced overall pesticide use (Unsworth, 2010). Despite the implementation of IPM and several other methods, the utilization of pesticides remains necessary to ensure an adequate supply of high-quality pest-free crops, prevent human diseases transmitted by pests, and positively impact public health (Fenik et al., 2011). However, pesticides can pose hazards to the health of the biota and the physical environment in several ways.

Several pesticides are biologically stable and bioaccumulate in the environment (Fenik et al., 2011). They can persist in the environment for years, leading to contamination through residues in air, water, and food (Mostafalou et al., 2013). Furthermore, a significant portion of pesticides applied to control weeds and pests is lost through various means, including off-target deposition and spray drift, as well as aerial sprays, surface runoff, and photodegradation, among others. This can have adverse effects on the species, communities, ecosystems, and human health (Hernández et al., 2013). Humans can be exposed to pesticides either directly or indirectly. Direct exposure occurs during pesticide usage in agriculture, public health programs, on livestock, and during fumigation processes. Indirect exposure occurs through the ingestion of contaminated food and water, as well as inhalation of pesticide droplets from drift (WHO, 2020; USEPA, 2022).

Pesticides exert their primary effects on the nervous and endocrine systems, making them potentially toxic to humans and capable of causing significant adverse health effects. Disruption of the endocrine system and the activation of inflammatory indicators can lead to the production of reactive oxygen species (ROS). These cause oxidative stress by interrupting the functioning of the endoplasmic reticulum and mitochondria within the cells. Even at minute concentrations, some pesticides can cause genetic disorders, epigenetic alterations, physiological alterations, and reduced lifespan in humans through various cellular processes (Poletta et al., 2009).

Scientific research has linked pesticide exposures to an amplified occurrence of various human diseases, including asthma, Alzheimer's, Parkinson's, amyotrophic lateral sclerosis (ALS), bronchitis, infertility, congenital disabilities, autism, diabetes, obesity, respiratory disorders, attention deficit hyperactivity disorder (ADHD), and organ system failures. These chemicals often threaten people with a higher chance of developing different kinds of cancers, such as non-Hodgkin lymphoma (NHL), leukemia, brain tumours, breast cancer, prostate cancer, lung cancer, stomach cancer, colorectal cancer, liver cancer, and urinary bladder cancer (WHO, 2020; USEPA, 2022). Children are predominantly susceptible to the detrimental effects of pesticides due to their unique physical constitution, physiology, and behaviour. Adverse health effects can even be manifested in individuals exposed to low levels of pesticides during early developmental stages.

Regardless of the need for pesticides, efforts must be made to strike a balance between environmental concerns and economic realities. The classification and understanding of pesticides based on their target pests and chemical structure play a critical role in managing their use and potential risks. To mitigate risks, the Codex Alimentarius committee and the Pesticide Data Program of the United States Environmental Protection Agency (USEPA) have established maximum residue limits for pesticides in edible food, which must be adhered to in order to prevent health hazards (WHO, 2020; USEPA, 2022).

This chapter provides comprehensive coverage of different types of insecticides, their utilization, targets, and modes of action. It also discusses the potential human exposure to pesticides and the subsequent impact on their health, with a particular emphasis on chronic effects, such as neurotoxicity, genotoxicity, carcinogenicity, and reproductive hazards. Health effects resulting from exposure to common groups of pesticides, such as organochlorines, organophosphates, carbamates, pyrethroids, fungicides, and herbicides, are also highlighted. Furthermore, the Chapter delves into the presence of pesticide residues in food, water, and crops, as well as their lethal consequences.

Types of Pesticides and Pesticide Formulations

Pesticides can be categorized in different ways. They can be categorized based on their chemical constituents, functional groups, mode of entry, mechanism of action, and the target organism. The majority of pesticides contain either organic (containing carbon) or inorganic (minerals such as copper, iron, lime, sulfur, etc.) components as active constituents. Organic pesticides exhibit hydrophobicity and are generally more complex than inorganic pesticides.

Organic pesticides can be further divided into natural pesticides, which are derived from naturally occurring sources, and synthetic pesticides, which are artificially produced through chemical synthesis. Pesticides can enter organisms through the integument, ingestion, drinking, or inhalation. Likewise, based on their mode of action, they can be categorized into neurotoxins, respiratory poisons, endocrine disruptors, physical poisons, protoplasmic poisons, chitin inhibitors, and others. According to the target organism, they can be categorized as insecticides, herbicides, fungicides, rodenticides, avicides, and more (EPA, 2004; Matthews et al., 2008). All these properties and categorizations of pesticides help to design and produce different kinds of formulations that can impart maximum effects for the utmost management of the target organism.

A pesticide's toxicity is determined by its effects on the target organism. The World Health Organization (WHO) categorizes pesticides into four types based on their LD50 level of toxicity (oral and dermal) to rats, i.e., the dosage that kills 50% of the rats (Table 1).

Table 1. Types of pesticides according to their toxicity, set by the World Health Organization (WHO)

Type of Pesticide	Toxicity Level	LD_{50} for the Rat (mg/kg Body Weight)		Examples
		Oral	Dermal	
Type Ia	Extremely hazardous	<5	<50	Parathion, Dieldrin
Type Ib	Highly hazardous	5–50	50–200	Eldrin, Dichlorvos
Type II	Moderately hazardous	50–2000	200–2000	DDT, Chlordane
Type III	Slightly hazardous	>2000	>2000	Malathion

Understanding the Adverse Effects of Synthetic Pesticides: A Comprehensive Overview

The effects and manifestations of pesticide noxiousness can range widely from mere minor skin irritation to stupor or even death. Different toxins belonging to various chemical classes or families impart different kinds of effects, causing the appearance of diverse symptoms. Individuals also exhibit varied sensitivity and responses to different pesticides. Some individuals may show no reaction, while others may develop mild to severe illness after exposure to pesticides. Hence, it is essential to recognize and understand the common signs and symptoms of pesticide poisoning. Pesticides and their effects on humans can be categorized into two types.

a) Topical effects of pesticide poisoning

These are the effects that generally develop at the site of pesticide contact due to either the irritant properties of the pesticide or an allergic response by the person. For example, dermatitis (inflammation of the skin) causes reddening of the skin or the development of rashes and/or blisters. An allergic response to pesticide exposure can cause coughing, sneezing, and sometimes severe respiratory symptoms, including wheezing and asthma, as well as red and itchy eyes, raw and scratchy nasal mucosa, and a raw and scratchy mouth lining and back of the throat. Many of these symptoms may persist, though they usually subside within a few minutes.

b) Systemic effects of pesticide poisoning

Systemic effects often occur at a location distant from the original point of contact/exposure. This is due to the absorption of pesticides in the body and their subsequent distribution throughout the body. Systemic effects often include physiological symptoms such as headaches, nausea, dizziness, vomiting, fatigue, and gastrointestinal issues. Advanced and acute pesticide poisoning can cause arrhythmia (irregular heart rate), dyspnea (shortness of breath or difficulty breathing), convulsions, and coma, eventually leading to death.

The poisoning effects of commonly used pesticides are discussed below.

Organochlorines

Organochlorine compounds, as the name suggests, are organic and belong to the group of chlorinated hydrocarbon derivatives. These are primarily utilized as insecticides against a variety of insect pests; typical examples include cyclodienes, HCH isomers (specifically the delta isomer, lindane), DDT and its analogues (e.g., DDE, methoxychlor, and dicofol), etc. Most of these pesticides, however, are residual, and they either do not degrade or degrade very slowly. The bioaccumulation and biomagnification of the toxins in the environment pose a significant public health concern.

Exposure to organochlorines occurs through ingestion of contaminated food or water, inhalation of vapour, and skin absorption. The lungs, gastrointestinal tract, and skin can absorb these compounds. Being liposoluble, their dermal penetration can be enhanced when used in lipid solvents, increasing the risk of intoxication, particularly among workers who use these products during crop cultivation without adopting adequate protection measures (Ellenhorn et al., 1988). Occupational and residential exposures are also possible. The consumption of contaminated food bioaccumulates the hazardous chemicals in the human body (Snedeker, 2001).

Intoxication with organochlorines causes the manifestation of various symptoms, including giddiness, headache, anorexia, sickness, vomiting, dermatitis, gastroenteritis, irritability, anxiety, gait syndromes, extreme sweating, altered reflexes, muscle weakness, shocks, spasms, mental confusion, seizures, coma, and even death. Studies have linked organochlorines to even cancer, wheezing, diabetes, and developmental abnormalities in children (Lee et al., 2006). The carcinogenic properties of organochlorines are associated with the polychlorocyclodiene compounds that undergo biotransformation to form epoxides.

The accumulation of organochlorines is attributed to their chemical and physical properties, such as polarity and solubility. These are lipophilic and can persist not only in the human body but also in the surrounding environment. Accordingly, several organochlorines have been banned by researchers and regulatory agencies (Barr and Needham, 2002). Due to the long half-life, the levels of these toxins in the serum can serve as a marker of exposure to them (Ellenhorn and Barceloux, 1988). Treatment of organochlorine toxicity requires maintaining the body's vital functions. Administration of diazepam, lorazepam, or phenobarbital through slow injection can help to prevent and manage seizures. Close monitoring of the airways is also important to avoid respiratory distress. In the case of myocardium-damaging arrhythmias, lidocaine is the line of treatment (Simpson and Schuman, 2002).

Organophosphates

Phosphorus-based organic compounds, known as organophosphorus compounds (OP), have been utilized as pesticides and war gases since their synthesis in 1937 (Barr et al., 2002). Based on their chemical structure, these compounds can be further categorized into phosphates, phosphorothioates, phosphoramidites, and phosphonates, which account for their distinctive characteristics (Afshari et al., 2004). Some examples of organophosphate pesticides are diazinon, malathion, and parathion (Ellenhorn and Barceloux, 1988).

Organophosphate poisoning is chiefly associated with the inhibition of the acetylcholinesterase (AChE) enzyme, which is present in both the CNS (Central Nervous System) and PNS (Peripheral Nervous System) and regulates the transmission of nerve impulses by the breakdown of the neurotransmitter acetylcholine. The organophosphates phosphorylate the serine hydroxyl group on AChE, inhibiting its action and resulting in an overabundance of acetylcholine, which leads to overstimulation of the nicotinic and muscarinic receptors.

OP compounds are readily absorbed either through the skin, conjunctiva, inhalation, or the gastrointestinal tract. The overstimulation of nicotinic and muscarinic receptors can lead to hypertension, sweating, tachycardia, psychosis, leukocytosis, respiratory distress, and nervous complications like hallucinations, flaccid paralysis, etc. Certain compounds, such as dichlofenthion, with high partition coefficients, can result in long-lasting symptoms (Davies et al., 1975). Ironically, the detoxification of these pesticides in the liver can sometimes produce metabolites that can impart higher toxicity (Costa, 2008). For example, the oxon form can bind to cholinesterase or undergo hydrolysis to form dialkyl phosphate and a hydrolyzed organic moiety, which can pose toxic effects (Barr and Needham, 2002).

Carbamates

Carbamate insecticides induce symptoms of cholinergic excess, manifested clinically, resembling the OP toxicity effects; however, the effects of carbamates are largely more reversible and less severe (Barr and Needham, 2002). The toxic mechanism underlying carbamate poisoning involves the carbamylation of the active site of the AChE enzyme, resulting in its inactivation (Ecobichon, 2001). Nevertheless, the carbamylated AChE can be reactivated through hydrolysis more rapidly compared to the phosphorylated enzyme, leading to the inhibition reversal usually within ≤30 min after exposure (Jokanovic, 2009). However, there have been reports of neuropathy following carbamate poisoning (Marrs and Maynard, 2013).

Like OP compounds, carbamates are also readily absorbed by organisms through the lungs, gastrointestinal tract, and skin. Fortuitously, the limited ability of carbamates to penetrate the blood-brain barrier results in minimal effects on brain cholinesterase activity, leading to fewer central nervous system symptoms compared to organophosphates. Additionally, the unprompted *in vivo* hydrolysis of the carbamate-cholinesterase complex contributes to the symptoms with short duration and less severity.

The primary symptoms of carbamate intoxication include constriction of the pupil (miosis), excess salivation, profuse sweating, bronchospasm, runny nose (rhinorrhea), behavioural alterations, stomach ache, nausea, diarrhea, urinary incontinence, dyspnea, low oxygen levels (hypoxemia), slow heart rate (bradycardia), pulmonary edema, respiratory failure, decrease in body temperature (hypothermia), lack of coordination, tingling, numbness, tremors, and seizures. Less commonly observed symptoms include rapid heart rate (tachycardia), muscle spasms, muscle weakness (including weakness in respiratory muscles), twitching, and paralysis (Rosman et al., 2009).

The treatment for carbamate intoxication usually involves the administration of atropine, followed by the potential use of furosemide, if necessary (Rosman et al., 2009). It is crucial to avoid the use of parasympathomimetic agents, as they can enhance the anticholinesterase activity.

Pyrethroids

Pyrethrins are natural compounds derived from the flowers of *Chrysanthemum cinerariaefolium* and consist of several active agents, with pyrethrins I and II being the most potent. However, these compounds are photolabile and degrade quickly under light. To overcome this limitation, pyrethroids, a synthetic derivative of these chemicals, were produced in ~1950, addressing some of the drawbacks associated with natural pyrethrins (USEPA, 2022).

Crude pyrethrum, which is derived from pyrethrins, can cause dermal and respiratory allergies. The allergic reactions, including contact dermatitis, rhinitis, and asthma, are likely attributed to the non-insecticidal components present in pyrethrum. These compounds possess an acid and an alcohol moiety linked by an ester bond. These exist as both the *trans* and *cis* stereoisomers, comprising a total of eight stereoisomers. These are further classified into Type I and Type II pyrethroids based on the presence or absence of a cyano (-CN) group in the alpha position (Zhang et al., 2010).

Pyrethroids are primarily absorbed through cutaneous routes or *via* inhalation and distribute rapidly within the body. They undergo biotransformation through either ester linkage hydrolysis by carboxylesterase enzymes or microsomal oxidation of the alcohol moiety by CYP450. Pyrethroids disturb the muscular system by interfering with the normal functioning of voltage-sensitive sodium channels. They bind to the α-subunit of the sodium channel, delaying its closure, which results in increased permeability of the cell membrane to Na^+ and a prolonged influx (tail current). It lowers the action potential threshold, resulting in repetitive discharge (Ray and Fry, 2006). At high concentrations, pyrethroids can block the generation of action potentials due to hyperexcitability, resulting in paralysis. They can also act on γ-aminobutyric acid-gated (GABA-gated) chloride channels, causing seizures and convulsions.

Exposure to crude pyrethrum can cause dermal and respiratory allergies (Reigart and Roberts, 1999; USEPA, 2022). Common toxic symptoms induced by pyrethroid poisoning include a tingling sensation on the face (facial paraesthesia), itchy and burning skin, nausea, and dizziness. Life-threatening symptoms include paralysis, convulsions, coma, pulmonary edema, and hemorrhage.

Other Pesticide Classes

Herbicides, such as chloroacetanilides, are commonly utilized in agriculture. Some chloroacetanilides, such as alachlor, acetochlor, metolachlor, and propachlor, have been identified as carcinogenic substances (Dearfield et al., 1999). The metabolism of chloroacetanilides is believed to involve conjugation with glutathione, as evidenced by the presence of glutathione-related metabolites in the urine of rats exposed to these herbicides (Ashby et al., 1996). Reports have suggested that these herbicides interfere with the biosynthetic processes of proteins, fatty acids, lipids, and flavonoids. However, due to predicted dissimilarities between humans and rats in terms of metabolism, concerns have been raised about their relevance in causing carcinogenicity in humans (Heydens et al., 2001). However, reports exist that demonstrate the acute oral intoxication of pretilachlor in humans and fish, causing clinical manifestations similar to those of organophosphate toxicity (Shilpakar et al., 2020; Ramesh et al., 2023).

Another noteworthy class of pesticides, the benzimidazoles, has been applied as veterinary medicines against helminths. These compounds bind to the free β-tubulin monomers at the colchicine-binding site, impeding the development of microtubules (Ermler et al., 2013). The toxicity of compounds has been elucidated on human RBCs by influencing the RBC acetylcholinesterase (Błaszczak-Świątkiewicz et al., 2016).

Likewise, each group of pesticides, although advantageous in various fields, has been found to impart toxicity to humans and other organisms with varying severity and effects. In the realm of new technologies, nano-pesticides are being investigated and formulated at a large scale. These encompass extremely small particles of an active ingredient or other formulated structures that possess beneficial or toxic properties (Bergeson et al., 2010). These formulations combine various surfactants, organic polymers, and inorganic metallic particles, such as silver and gold, in the nanometer size range (Sekhon, 2014). Such technologies offer several advantages when using these pesticides, including enhanced efficacy, durability, and the requirement of smaller quantities of active ingredients (Kookana et al., 2014).

Impact of Pesticides on the Soil, Plants, and Aquatic Systems

The agricultural use of pesticides can pollute soils in several ways, harming the soil flora and fauna, depending on their features such as toxicity, residual nature, and bioaccumulation. In European countries, an assessment of 317 agricultural topsoil samples has shown the existence of one or more pesticide residues (Silva et al., 2019). The most common cause of soil contamination is the haphazard use of pesticides without adequate knowledge of the frequency, means, and concentration to be applied. The extensive and arbitrary use of pesticides on various crop species has led to significant amounts of pesticide runoff into water bodies, as well as their adsorption, desorption, and degradation while seeping into the soil. All these phenomena have harmed our ecosystem, plants, and beneficial soil biota, such as microorganisms and small animals. Sadly, the bioavailability of pesticides in the soil enables them to enter the food web through uptake, diffusion, and dispersion, thereby magnifying their effects at each level and affecting multiple organisms. Apart from these effects, plants take up these pesticides from the soil along with the water. Pesticides translocate through the vascular system, affecting vital processes in every part, such as stomatal movement, photosynthesis, and respiration.

Persistent and current-use pesticides can enter water bodies through various modes, primarily through runoff *via* atmospheric precipitation and the release of waste from pesticide manufacturing industries into water bodies. Studies have shown the presence of high concentrations of certain pesticides, including atrazine, chlorpyrifos, and napropamide, in the aquatic system of Shanghai, China (Chen et al., 2020). Moreover, pesticides can contaminate groundwater through agricultural runoff and leaching. It deteriorates the water quality, making it unsafe for human consumption. Moreover, the use of contaminated water can be hazardous to both flora and fauna. Additionally, the pesticides accumulate and biomagnify in the food chain, affecting other organisms and, eventually, human health.

The presence of pesticides in water bodies contaminates aquatic ecosystems, harming aquatic flora and fauna (Woodrow et al., 2018). Hematological parameters have been reported in fish exposed to pesticides. Abnormal counts of blood cells, altered serum levels, and abnormalities in vital organs, such as the liver, kidneys, muscles, gut, and brain, have been demonstrated (Tahir et al., 2021). Along with physiological and behavioural alterations, such as developmental abnormalities, decreased reproductive capacity, hyperexcitability, increased respiration, irregular activities, and even variations in body colour have been reported (Dawar et al., 2016).

Impact of Pesticides on Human Health

Humans can be exposed to pesticides through various routes, both actively through occupational exposure and passively through non-occupational exposure. Occupational exposure to pesticides can occur in humans during various agricultural and industrial processes, such as manufacturing, transportation, and pesticide application, among others. In Pakistan, 800 workers developed toxic symptoms during the malathion spray for malaria vector control (USAID, 2007). In the agricultural industry, workers may unknowingly bring pesticide-contaminated clothes and equipment home, which has been associated with an increased risk of cancer development in their children. The adverse effects of pesticides, along with some heavy metals, on human health have been linked to various living systems.

Non-occupational exposure to pesticides can occur through the ingestion of pesticide residues in contaminated food and water, inhalation of pesticide droplets in the air that drift from the point of release or fumigation, and dermal exposure to residual indoor sprays or outdoor fumigation with insecticides used for domestic or public health purposes. Additionally, exposures can be achieved while treating the ectoparasites in pets, such as fleas, ticks, or mites. Children, particularly, are vulnerable to such exposures.

Dermal exposure, which occurs through contact with the skin, is the most common route of pesticide entry. This can occur through incidents such as splashes, spills, or spray drifts during pesticide-related activities, including mixing, loading, applying, and disposing. Even cleaning of application equipment, particularly without the use of appropriate protective clothing, is a significant source of exposure. The absorption of pesticides is correlated with the amount/concentration of pesticide, duration of exposure, and environmental factors, such as temperature and humidity. Liquid formulations, such as emulsifiable concentrates (EC), are more easily absorbed through the skin relative to solid formulations like powders, dusts, and granules. Certain body areas, such as the groin, eyes, and ear canal, have higher absorption rates. Accidental ingestion of pesticides can occur when individuals consume food or drinks contaminated with pesticide residues from unlabeled containers used for pesticide storage or during the handling of pesticides. Apart from this, the application of fumigants or other pesticide formulations leaves fine droplets in the air, which humans inhale.

Once pesticides enter the human body, they are transported throughout the body via blood circulation, metabolized, absorbed, and eventually excreted through urine, skin, and exhalation. All these processes and pathways contribute to the noxious effects of pesticides. Persistent organic pollutants (POPs), being lipophilic, accumulate in human adipose tissue, breast milk, and maternal blood, as well as the placenta, causing significant harm to the body and imparting lethal effects of varying severity (Pesticides formulations, 2020). These effects may be manifested as genotoxicity, cytotoxicity, or hormonal changes, all of which lead to carcinogenic effects. The pesticide-associated disorders include reproductive disorders and congenital abnormalities (Castillo-Cadena et al., 2017; García et al., 2017), various types of cancer (Mostafalou and Abdollahi, 2017), depression, nervousness, Parkinson's and Alzheimer's diseases (Sharma et al., 2020), which may eventually lead to the death of an individual. Let us examine some of the mechanisms by which pesticides can cause lethal effects in the human body.

Reactive Oxygen Species

The toxic effects of numerous pesticides are connected with the production of free radicals (Reactive oxygen species or ROS) that possess one or more unpaired electrons. They possess the potential to cause peroxidation of cellular lipid layers and contribute to the body's antioxidant defense mechanisms. Hence, they are not only lethal themselves but also contribute to the pathophysiology of various diseases due to oxidative stress created by their imbalance in the body.

Prolonged occupational exposure to various heavy metals, such as cadmium and lead, is another underlying cause of oxidative stress induction, resulting in adverse effects in biological systems. The involvement of oxidative stress as a significant mechanism has been evidenced in Alzheimer's, Parkinson's, cataracts, neoplastic diseases, atherosclerosis, diabetes, aging, chronic inflammation of the GI tract, asthma, etc. A comprehensive list of diseases in which oxidative stress is possibly implicated in the pathophysiology of various diseases is presented.

Over the past decade, pesticide-induced oxidative stress has garnered considerable attention in toxicological research as a potential phenomenon of toxicity. Numerous studies have investigated the association between oxidative stress and humans or animals exposed to various chemical agents. To comprehend the precise nature of oxidative stress, it is necessary to elucidate the principles of free radical production and the genesis of the body's standard defense system.

The human body possesses various mechanisms to counter and combat the detrimental effects of free radicals. The primary defence mechanism employed is through the utilization of antioxidant agents, which delay or inhibit the oxidative damage to specific targeted molecules. These are stable compounds and can neutralize free radicals by donating electrons. They exert their beneficial effects by modulating the levels of harmful radicals and regulating various redox signalling pathways. Over time, numerous compounds with antioxidant activity have been identified, which are classified into two primary systems in humans.

The first line of defense is an enzymatic system, which is responsible for maintaining pools of antioxidant vitamins, such as β-carotene, tocopherol, and ascorbic acid. This system can neutralize all free radicals unless oxidative stress exceeds the system's capacity. In that case, a secondary line of defense gets activated, which utilizes vitamins to scavenge and quench free radicals. However, they become oxidized and inactivated in the process. Each of these antioxidant nutrients exhibits specific activity and frequently works synergistically to augment the overall antioxidant capacity of the body (Halliwell and Gutteridge, 1990).

However, upon ingestion, antioxidants undergo extensive metabolism, resulting in a diverse array of bioactive metabolites. These metabolites contribute to the diverse bioactivities exhibited by antioxidants. The failure of the body's natural antioxidant system to cope with the free radicals can cause cellular damage through a sequence of events. The damage includes the destruction of cell membranes, inhibition of important enzymes, disruption of vital cellular processes, hindrance of normal cell division, DNA damage, and impaired energy generation. Consequently, oxidative stress has been implicated in the pathophysiology of metabolic disorders, chronic diseases, and even cancer (Halliwell and Gutterdge, 1990; Sies and Stan, 1998).

Understanding the mechanisms of action and metabolites of antioxidants is crucial for improving their therapeutic applications. An in-depth exploration of the metabolite fingerprint, known as the scavengome, which represents the array of metabolites involved in free radical scavenging, can enhance our understanding of the bioactivity of antioxidants (Sies and Stan, 1998).

Endocrine Disruption

The endocrine system is a complex network of ductless glands, chemical messengers, and receptors that play a vital role in regulating various bodily functions, including reproduction, metabolism, growth and development, stress response, and homeostasis (the maintenance of balance and stability). It acts as a messenger that serves as the communication and regulatory link between the nervous system and bodily functions, maintaining life processes (Lintelmann et al., 2003). Disruptions or disturbances in the endocrine system can have significant impacts on the overall development of organisms. The exogenous substances that modify hormonal actions, leading to adverse health effects in organisms, are termed endocrine-disrupting chemicals (EDCs), such as environmental estrogens (xenoestrogens), androgenic compounds, and thyroid-activating chemicals (Mnif et al., 2011). EDCs can interfere with hormone receptors, either by imitating the action of natural hormones, blocking hormone receptors, or interacting with multiple receptors (Endocrine disrupters website, 2011), causing hormonal disorders across different animal classes, including mammals.

Cancer Risks

Occupational exposure to agricultural pesticides has been reported to increase cancer risks in farmers and associated individuals due to DNA damage resulting in mutations, chromosomal breaks, double-stranded chromosomes, and the formation of DNA adducts.

Several types of cancers, such as breast cancer, multiple myeloma, non-melanoma skin cancer, and colorectal cancer, have been documented due to exposure to herbicides in farmers (Salerno et al., 2016). Notably, males were found to develop more incidences of multiple myeloma, while females developed melanoma (Lemarchand et al., 2017). Apart from these two primary cancers, bladder cancer (Boada et al., 2016), non-Hodgkin Lymphoma (NHL) (Zakerinia et al., 2012), and prostate cancer (Lemarchand et al., 2016) have been reported in pesticide workers exposed to organochlorines and organophosphates. A noteworthy danger of tumor development in the CNS has also been identified in farmers and workers who frequently apply pesticides in the fields (Piel et al., 2017).

Assessment and Mitigation of Pesticide-Induced Risks

Understanding the various routes and pathways of pesticide exposure is crucial in assessing and mitigating the potential health risks associated with these chemicals (Pimental, 2009; Pesticides Formulations, 2020). Detoxification measures following instances of poisoning are of utmost importance, regardless of whether the exposure was accidental, intentional, or occupational. The identification of poisoning becomes easier if the individual possesses knowledge regarding the specific pesticide to which they were exposed or when the poisoning symptoms align with the typical indicators. However, in cases of generalized symptoms, the diagnosis of poisoning is often uncertain and unclear. Therefore, in addition to implementing procedures to mitigate the contamination, it is essential to investigate the incident and gather relevant information on patient care. The interviews of the individuals exposed can ascertain the method of exposure to the toxin and assess the potential for concurrent intoxication with other poisons (Simpson and Schuman, 2002). Alongside, the mandatory analysis techniques can detect the severity of pesticides.

The alleviation of poisoning effects must be accompanied by the monitoring and maintenance of vital signs, as well as the administration of antidotes to prevent the emergence of new contamination cases. Such patients should be kept isolated; otherwise, the medical professionals and other patients, if in the same ward as the pesticide-contaminated persons, must take protective measures until the completion of decontamination and treatment procedures (Simpson and Schuman, 2002).

Decontamination of individuals poisoned through the gastrointestinal tract can be done in various ways. Potentially life-threatening cases can be managed by gastric lavage. The cathartic method used to stimulate bowel movements and facilitate the excretion of pesticides is not appropriate in cases of diarrhea. An alternative approach involves the administration of adsorbents, such as charcoal. However, it is unable to adsorb all pesticides, which bind with the toxin and form a stable compound. The GI tract cannot absorb this compound, resulting in its elimination through fecal excretion. This procedure is often conducted along with the cathartic method. Finally, the administration of ipecac syrup, a plant product, can trigger vomiting. However, this procedure is not advised in cases involving the ingestion of hydrocarbons or corrosives (Reigart and Roberts, 1999; Simpson and Schuman, 2002).

In instances of dermal exposure, the patient must be placed under a shower instantly to remove the chemicals from every body part using soap and water.

In case of contact through the ocular route, the affected eye should be rinsed immediately with copious amounts of clean water. The patient's clothing and other belongings should be discarded. In case of significant contamination, it is crucial to decontaminate each individual involved in the emergency response system (Reigart and Roberts, 1999; Simpson and Schuman, 2002).

Generally, all the treatments mentioned above aim to counteract the mechanisms of action of toxins, with numerous opportunities for intervention. Given the vast array of pesticide compositions available, the clinical profile and treatment associated with major poisoning incidents, in terms of quantity and severity, need to be focused on.

Integrated Pest Management and Technologies for Safer Insecticides

Integrated Pest Management (IPM) is a highly effective and environmentally conscious approach to pest control. It relies on a synergistic and harmonious combination of diverse practical measures in order to keep our environment and biodiversity safe. IPM programs incorporate up-to-date and comprehensive information and data on the biology of pests and their interactions with their environment. This valuable knowledge, combined with a range of available pest control methods, enables the management of pest populations below their threshold limits in the most economically viable manner, while mitigating the risks to people, their belongings, and the environment (Abid et al., 2021).

IPM, which can be applied in both agricultural and non-agricultural situations, encompasses a wide range of management options. These include cultural methods, the use of resistant plant varieties, physical and mechanical measures, the use of traps, biological control agents, and the cautious application of pesticides. The process of IPM involves a comprehensive, four-tiered approach:

a) Setting Action Thresholds: IPM establishes specific thresholds before taking any pest control measures. The limits are set at which control measures of pest populations are necessary to implement. A single pest sighting may not always warrant intervention; instead, the economic threat posed by pest levels guides decision-making.

b) Monitoring and Identifying Pests: IPM emphasizes the importance of accurately monitoring and identifying pests. Not all pests require control, as some are harmless or even beneficial to the environment. Proper monitoring ensures that pesticides are used only when necessary and that appropriate control measures are selected.
c) Prevention: The first line of defence in IPM is prevention, aimed at managing crops, fields, or indoor spaces to deter pests from becoming a significant threat. Cultural methods, such as crop rotation, using pest-resistant plant varieties, and employing pest-free rootstock, can be highly effective, cost-efficient, and pose minimal risks to the environment and human health.
d) Control: When monitoring and preventive measures are insufficient and pest control becomes necessary, IPM evaluates suitable methods based on their effectiveness and risk levels. Less risky and targeted pest controls are prioritized, such as the use of allelochemicals to disrupt mating, the setup of mechanical traps, and weeding. Only if these approaches prove ineffective will more extensive measures, such as targeted pesticide application, be considered as a last resort.

IPM's effectiveness is demonstrated through various projects worldwide. A total of 85 IPM projects implemented over two decades were evaluated across 24 countries in Asia and Africa. The results showed an average crop yield increase of 40.9%, accompanied by a 30.7% decline in pesticide use compared to baseline levels. Remarkably, 30% of crop combinations transitioned to zero pesticide use, indicating the significant potential of IPM to reduce reliance on pesticides (Pretty and Bharucha, 2015).

Despite its effectiveness, IPM faces challenges, such as limited policy support and opposition from the pesticide industry. As pests, diseases, and weeds evolve and migrate, the implementation of IPM remains an ongoing challenge. However, as more evidence supports its positive impacts and sustainability, the adoption of IPM continues to grow, offering a promising pathway to more responsible pest management practices (Deguine et al., 2021).

Conclusion

In conclusion, the use of insecticides has become indispensable in modern life for protecting various settings, such as homes, farms, and gardens, from destructive pests and the transmission of infectious diseases. However, the continuous development of novel insecticide formulations is necessary due to the rise of insect resistance, environmental concerns, and the association of insecticides with adverse human health effects. The widespread application of residual and non-degradable insecticides has led to the contamination of air, water, and soil ecosystems, resulting in the decline of beneficial organisms such as predators, pollinators, and earthworms, as well as alterations in soil microbial communities. Additionally, the persistence of insecticides has led to their accumulation in food chains, reaching humans and other mammals. These consequences highlight the detrimental impact of indiscriminate insecticide use on various elements of the ecosystem. Therefore, it is crucial to use insecticides responsibly to safeguard the environment and mitigate potential health risks. This chapter examined the diverse effects of insecticides on ecosystems, plants, and human health. Various aspects of insecticide action and effects were discussed at length. Furthermore, the Integrated Pest Management (IPM) approach, which combines multiple control measures such as cultural practices, resistant genotypes, physical and mechanical control, and judicious use of insecticides, has been highlighted as a means to mitigate the harmful effects of insecticides by reducing their quantity and frequency of application. The potential of leading-edge techniques in developing safer insecticides has also been highlighted. It is recommended to prioritize community development and extension initiatives that can inculcate knowledge and inspire farmers to adopt advanced IPM strategies.

References

Abdollahi M, Bahreini-Moghadam A, Emami B, Fooladian F, Zafari K. Increasing intracellular cAMP and cGMP inhibits cadmium-induced oxidative stress in rat submandibular saliva. *Comparative Biochemistry and Physiology Part-C: Toxicology* (2003) 135(3): 331–336.

Abid I, Laghfiri M, Bouamri R, Aleya L, Bourioug M. Integrated pest management (IPM) for *Ectomyelois ceratoniae* on date palm. *Current Opinion in Environment Science and Health* (2021) 19: p.100219.

Afshari R, Majdzadeh R, Balali-Mood M. Pattern of acute poisonings in Mashhad, Iran 1993–2000. *Journal of Toxicology: Clinical Toxicology* (2004) 42(7):965–975.

Alengebawy A, Abdelkhalek ST, Qureshi SR, Wang M-Q. Heavy metals and pesticides toxicity in agricultural soil and plants: ecological risks and human health implications. *Toxics* (2021) 9(3):42. https://doi.org/10.3390/toxics9030042.

Ashby J, Kier L, Wilson AG, Green T, Lefevre PA, Tinwell H, Willis GA, Heydens WF, Clapp MJ. Evaluation of the potential carcinogenicity and genetic toxicity to humans of the herbicide acetochlor. *Human & Experimental Toxicology* (1996) 15(9): 702-735.

Barr DB, Needham LL. Analytical methods for biological monitoring of exposure to pesticides: a review. *Journal of Chromatography B* (2002) 778(1-2):5-29.

Bergeson LL. Nanosilver: US EPA's pesticide office considers how best to proceed. *Environment Quality Management* (2010) 19(3): 79–85. http://onlinelibrary.wiley.com/doi/10.1002/tqem.20255/pdf..

Bilal M, Iqbal Hafiz MN, Barcelo D. Persistence of pesticides-based contaminants in the environment and their effective degradation using laccaseassisted biocatalytic systems. *Science of Total Environment* (2019) 695:133896.

Błaszczak-Świątkiewicz K, Sikora J, Szymański J, Danilewicz M, Mikiciuk-Olasik E. Biological evaluation of the toxicity and the cell cycle interruption by some benzimidazole derivatives. *Tumour Biology* (2016) 37(8):11135-11145.

Boada LD, Henriquez-Hernandez LA, Zumbado M, Almeida-Gonzalez M, Alvarez-Leon EE, Navarro P, Luzardo OP. Organochlorine pesticides exposure and bladder cancer: evaluation from a gene-environment perspective in a hospital-based case-control study in the Canary Islands (Spain). *Journal of Agromedicine* (2016) 21:34–42.

Butterfield DA, Lauderback CM. Lipid peroxidation and protein oxidation in Alzheimer's disease brain: potential causes and consequences involving amyloid beta-peptide-associated free radical oxidative stress. *Free Radical Biology and Medicine* (2002) 32(11):1050–1060.

Butterfield DA. Amyloid beta-peptide (1–42) induced oxidative stress and neurotoxicity: implications for neurodegeneration in Alzheimer's disease brain. A review. *Free Radical Research* (2002) 36(12): 1307–1313.

Castillo-Cadena J, Mejia-Sanchez F, López-Arriaga JA. Congenital malformations according to etiology in newborns from floricultural zone of Mexico state. *Environmental Science and Pollution Research International* (2017) 2:7662–7667.

Chen C, Zou W, Cui G, Tian J, Wang Y, Ma L. Ecological risk assessment of current-use pesticides in an aquatic system of Shanghai, China. *Chemosphere* (2020). 257:127222.

Costa LG. Toxic Effects of Pesticides. In: Klaassen CD. (ed). *Cassarett and Doull's Toxicology: The Basic Science of Poisons*. 7 Ed, McGraw-Hill, New York (2008) 8:883-930.

Davies JE, Barquet A, Freed VH, Haque R, Morgade C, Sonneborn RE, Vaclavek C. Human pesticide poisonings by a fat soluble organophosphate insecticide. *Archives of Environmental Health* (1975) 30(12): 608-613.

Dawar FU, Zuberi A, Azizullah A, Khattak MNK. Effects of cypermethrin on survival, morphological and biochemical aspects of rohu (*Labeo rohita*) during early development. *Chemosphere* (2016) 144: 697–705.

Dearfield KL, McCarroll NE, Protzel A, Stack HF, Jackson MA, Waters M. D. A survey of EPA/OPP and open literature on selected pesticide chemicals. II. Mutagenicity and carcinogenicity of selected chloroacetanilides and related compounds. *Mutation Research* (1999) 443(1-2): 183-221.

Deguine JP, Aubertot JN, Flor RJ, Lescourret F, Wyckhuys KA, Ratnadass A. Integrated pest management: good intentions, hard realities. A review. *Agronomy for Sustainable Development* (2021) 41(3): p.38.

Ecobichon DJ. Carbamate insecticides. In: Krieger R (ed). *Handbook of Pesticide Toxicology*. San Diego, CA: Academic Press (2001): 1087-1106.

Ellenhorn MJ, Barceloux DG. Medical Toxicology. In: *Diagnosis and Treatment of Human Poisoning*. Ed. Elsevier, New York (1988): 1512.

EPA - Environmental Protection Agency. Overview of the Ecological Risk Assessment Process in the Office of Pesticide Programs, U.S. EPA (2004) http://www.epa.gov/espp/consultation/ecorisk-overview.pdf (Accessed 18 June 2023).

Ermler S, Scholze M, Kortenkamp A. Seven benzimidazole pesticides combined at sub-threshold levels induce micronuclei in vitro. *Mutagenesis* (2013) 28(4): 417-426.

Fenik J, Tankiewicz M, Biziuk M. Properties and determination of pesticides in fruits and vegetables. *TrAC* (2011) 30 (6): 814-826.

García J, Ventura MI, Requena M, Hernández AF, Parrón T, Alarcón R. Association of reproductive disorders and male congenital anomalies with environmental exposure to endocrine active pesticides. *Reproductive Toxicology* (2017) 71:95–100.

Gupta PK. Toxicity of herbicides. In: Gupta RC (Ed). *Veterinary Toxicology: Basic and Clinic Principles*. USA: Elsevier (2007):567-586.

Halliwell B, Gutterdge JMC. The antioxidants of human extracellular fluids. *Archives of Biochemistry and Biophysics* (1990) 280(1): 1–8.

Hernández AF, Parrón T, Tsatsakis AM, Requena M, Alarcón R, López-Guarnido O. Toxic effects of pesticide mixtures at a molecular level: Their relevance to human Health. *Toxicology* (2013) 307: 136-145.

Heydens WF, Lamb IC, Wilson AGE. Chloroacetanilides. In: Krieger R (ed). *Handbook of Pesticide Toxicology*. San Diego, CA: Academic Press (2001): 1543–1558.

Jokanovic M. Medical treatment of acute poisoning with organophosphorus and carbamate pesticides. *Toxicology Letters* (2009) 90(2):107-115.

Kookana RS, Boxal AB, Reeves PT, Ashauer R, Chaudhry Q, Cornelis G, Fernandes TF, Gan J, Kah M, Lynch I, Ranville J, Sinclair C, Spurgeon D, Tiede K, Van Den Brink PJ. Nanopesticides: guiding principles for regulatory evaluation of environmental risks. *Journal of Agriculture and Food Chemistry* (2014) 62(19): 4227-4240.

Lee DH, Lee IK, Song K, Steffes M, Toscano W, Baker BA, Jacobs DR. A strong dose-response relation between serum concentrations of persistent organic pollutants and diabetes. *Diabetes Care* (2006) 29(7): 1638-1644.

Lemarchand C, Tual S, Boulanger M, Levêque-Morlais N, Perrier S, Chin B, Guizard AV, Velten M, Rigaud E, Baldi I, Lebailly P. Prostate cancer risk among French farmers in the AGRICAN cohort. *Scandinavian Journal of Work, Environment and Health* (2016) 144–152.

Lemarchand C, Tual S, Levêque-Morlais N, Perrier S, Belot A, Velten M, Guizard AV, Marcotullio E, Monnereau A, Chin B, Baldi I, Lebailly P, the AGRICAN group. Cancer incidence in the AGRICAN cohort study (2005–2011). *Cancer Epidemiology* (2017) 49:175–185

Levengood JM, Beasley VR. Principles of ecotoxicology. In: Gupta RC (ed.) Veterinary toxicology: basic and clinical principles. Academic press, *Amterdan* (2007): 689-708.

Lintelmann J, Katayama A, Kurihara N, Shore L, Wenzel A. Endocrine disruptors in the environment. *Pure and Applied Chemistry* (2003) 75(5):631-681.

Marrs TC, Maynard RL. Neurotransmission system as targets for toxicants: a review. *Cell Biology and Toxicology* (2013) 29: 381-396.

Matisova E, Hrouzkov S. Endocrine disrupting pesticides. In: *Pesticides - Advances in Chemical and Botanical Pesticides*. InTech (2012). Available from: http://dx.doi.org/10.5772/46226.

Matthews G, Bateman R, Miller P. Book Pesticide application methods. John Wiley & Sons. (2008) Available from http://onlinelibrary.wiley.com/book/10.1002/9780470760130 (Accessed 28th July 2023).

Mnif W, Hassine AIH, Bouaziz A, Bartegi A, Thomas O, Roig B. Effect of endocrine disruptor pesticides: a review. *International Journal of Environment Research and Public Health* (2011) 8(6): 2265-2303.

Mostafalou S, Abdollahi M. Pesticides and human chronic diseases: Evidences, mechanisms, and perspectives. *Toxicology and Applied Pharmacology* (2013) 268(2): 157-177.

Mostafalou S, Abdollahi M. Pesticides: an update of human exposure and toxicity. *Archives of Toxicology* (2017) 91:549–599.

Pathak VM, Verma VK, Rawat BS, Kaur B, Babu N, Sharma A, Dewali S, Yadav M, Kumari R, Singh S, Mohapatra A, Pandey V, Rana N and Cunill JM. Current status of pesticide effects on environment, human health and it's eco-friendly management as bioremediation: A comprehensive review. *Frontiers in Microbiology* (2022) 13:962619.

Pedroso TMA, Benvindo-Souza M, de Araújo Nascimento F, Woch J, Dos Reis FG, de Melo e Silva D. Cancer and occupational exposure to pesticides: a bibliometric study of the past 10 years. *Environmental Science and Pollution Research* (2022) 1-12.

Piel C, Pouchieu C, Tual S, Migault L, Lemarchand C, Carles C, Boulanger M, Gruber A, Rondeau V, Marcotullio E, Lebailly P, Baldi I. Central nervous system tumors and agricultural exposures in the prospective cohort AGRICAN. *International Journal of Cancer* (2017) 141:1771–1782.

Pimentel D. *Integrated Pest Management*. In: Peshin, Rajinder, Dhawan AK, editor. 1st ed. Springer (2009).

Poletta GL, Larriera A, Kleinsorge E, Mudry MD. Genotoxicity of the herbicide formulation Roundup® (glyphosate) in broad-snouted caiman (Caiman latirostris) evidenced by the Comet assay and the Micronucleus test. *Mutation Research* (2009) 672(2): 95-102.

Pretty J, Pervez Bharucha, Z. Integrated pest management for sustainable intensification of agriculture in Asia and Africa. *Insects* (2015) 6(1):152-182.

Ramesh M, Priyanka M, Tamilarasi R, Poopal RK, Ren Z, Li B. Short-term toxicity of chloroacetanilide herbicide on non-target organism: estimation of median-lethal concentration, hematological, biochemical, ion regulation, and antioxidants. *Comparative Clinical Pathology* (2023) 32(1):125-137.

Ratcliffe D. Decreases in eggshell weight in certain birds of prey. *Nature* (1967) 215: 208-210.

Ray DE, Fry JR. A reassessment of the neurotoxicity of pyrethroid insecticides. *Pharmacology Therapy* (2006) 111(1): 174-193.

Reigart JR, Roberts JR. *Recognition and Management of Pesticide Poisonings*. NPIC (1999): 119-132. Retrieved from http://www.epa.gov/oppfead1/safety/healthcare/handbook/handbook.pdf (Accessed 28 July 2023).

Reigart JR, Roberts JR. *Recognition and Management of Pesticide Poisonings*. 5th Ed. U.S. EPA (1999) http://www.epa.gov/oppfead1/safety/healthcare/handbook/handbook.pdf (Accessed 28 July 2023).

Rosman Y, Makarovsky I, Bentur Y, Shrot S, Dushnistky T, Krivoy A. Carbamate poisoning: treatment recommendations in the setting of a mass casualties event. *American Journal of Emergence Medicine* 2009; 27[9]:1117-1124.

Salerno C, Carcagnì A, Sacco S, Palin LA, Vanhaecht K, Panella M, Guido D. An Italian population-based case-control study on the association between farming and cancer: Are pesticides a plausible risk factor? *Archives of Environmental & Occupational Health* (2016) 71:147–156.

Sekhon BS. Nanotechnology in agri-food production: an overview. *Nanotechnology, Science & Applications* (2014) 7: 31-53.

Sharma A, Shukla A, Attri K, Kumar M, Kumar P, Suttee A, Singh G, Barnwal RP, Singla N. Global trends in pesticides: a looming threat and viable alternatives. *Ecotoxicology and Environmental Safety* (2020) 201:110812.

Shilpakar O, Karki B, Rajbhandari B. Pretilachlor poisoning: A rare case of a herbicide masquerading as organophosphate toxicity. *Clinical Case Reports* (2020) 8(12):3506-3508.

Sies H, Stan LW. Vitamins E and C, carotene and other carotenoids as antioxidants. *American Journal of Clinical Nutrition* (1998) 62(6):1315S–1321S.

Silva V, Mol HG, Zomer P, Tienstra M, Ritsema CJ, Geissen V. Pesticide residues in European agricultural soils–A hidden reality unfolded. *Science of the Total Environment* (2019) 653:1532–1545.

Simpson Jr WM, Schuman SH. Recognition and management of acute pesticide poisoning. *American Family Physician* (2002) 65(8): 1599-1604.

Snedeker SM. Pesticides and breast cancer risk: a review of DDT, DDE and dieldrin. *Environmental Health Perspectives* (2001) 109 (suppl 1): 35-47.

Tahir R, Ghaffar A, Abbas G, Turabi TH, Kausar S. Pesticide induced hematological, biochemical and genotoxic changes in fish: a review. *Agrobiological Records* (2021) 3:41 57.

United States Environmental Protection Agency. (2022). *Human Health Issues Related to Pesticides.* Retrieved from https://www.epa.gov/pesticide-science-and-assessing-pesticide-risks/human-health-issues-related-pesticides (Accessed 18 June 2023).

University of Kentucky. PESTICIDE FORMULATIONS. *Kentucky Pesticide Safety Education Program* [Internet] (2020) Available from: https://www.uky.edu/Ag/Entomology/PSEP/3formulations.html.

Unsworth J. *History of pesticide use*. IUPAC (2010). Available from: http://agrochemicals.iupac.org/index.php?option=com_sobi2&sobi2Task=sobi2Details&catid=3&sobi2Id=31. Accessed June 18, 2023.

USEPA - United States of Environmental Protection Agency. About pesticides. US EPA. http://www.epa.gov/pesticides/about/index.htm (Accessed 17 June 2023).

Woodrow JE, Gibson KA, Seiber JN. Pesticides and related toxicants in the atmosphere. *Rev Environ Conta Toxicol* (2018) 247, 147–196.

World Health Organization. (2020). *Chemical safety: Pesticides*. Retrieved from https://www.who.int/news-room/questions-and-answers/item/chemical-safety-pesticides (Accessed 18 June 2023).

Zakerinia M, Namdari M, Amirghofran S. The relationship between exposure to pesticides and the occurrence of lymphoid neoplasm. *Iran Red Crescent Medical Journal* (2012) 14:337.

Zarkovic N. 4-Hydroxynonenal as a bioactive marker of pathophysiological processes. *Molecular Aspects of Medicine* (2003) 24(4-5): 281–291.

Zhang Q, Zhang W, Wang X, Li P. Immunoassay Development for the Class-Specific Assay for Types I and II Pyrethroid Insecticides in Water Samples. *Molecules* (2010) 15(1):164-177.

About the Editors

Dr Archana Anokhe is a Scientist at the Indian council of Agricultural Research- Directorate of weed Research, Jabalpur, under the Ministry of Agriculture and Farmers Welfare. She has also served as a Visiting Scholar at the Agricultural Research Organisation (ARO), Israel. She has an exceptional academic record, having qualified for the Junior Research Fellowship (AIR-13), Senior Research Fellowship (AIR-4), and Agricultural Research Services (AIR-8). Her achievements include the Women Scientist Award (2022), Young Scientist Award (2021), Best PhD Thesis Award (2020), Best Student Award (2019), and several Best Oral Presentation Awards at the National and International Conferences. She was also a recipient of the National Talent Scholarship, UGC National Fellowship, IARI Fellowship, and MASHAV International Fellowship. In addition to her post-graduation and PhD in Entomology, she holds a Post-Graduate Diploma in Education Technology Management, a Post-Graduation Diploma in Dr Ambedkar Thoughts, and a Master of Arts in English.

Dr Anokhe has been actively engaged in teaching courses such as Insect Pathology, Host Plant Resistance, Advances in Host Plant Resistance, and Pests of Horticultural and Plantation Crops. She has also guided 5 M.Sc. and Ph.D. students. She has published more than 50 research papers in national and international journals, along with over 100 popular articles, blogs, and technical bulletins. She serves as the Associate Editor of the *Indian Journal of Entomology*, *Indian Entomologist* magazine, and *Agro-Titan* magazine, and is a member of several institutional committees. She has also been featured in interviews on Doordarshan and All India Radio. Her research specialisation lies in insect pathology and virus–vector interaction studies, and Invasive weed management.

Beyond academia and research, Dr. Archana Anokhe has made significant contributions to managing invasive weeds that threaten agriculture, biodiversity, and rural livelihoods—especially *Parthenium hysterophorus* and *Salvinia molesta*. She has played a leading role in organizing the nationwide Annual Parthenium Awareness Week programme, mobilizing more than 100 ICAR institutes, 730 Krishi Vigyan Kendras, and 75 State Agricultural Universities to promote eco-friendly management practices. A strong proponent of biological control, she coordinates the free large-scale distribution of the Mexican beetle (*Zygogramma bicolorata*) to farmers; in 2024–25 alone, over 3 lakh beetles were distributed.

Alongside terrestrial weed management, Dr. Archana Anokhe has also contributed significantly to aquatic ecosystem restoration. In 2023–24, she successfully managed *Salvinia molesta* infestations in Murwari (25 acres) and Gidurha (15 acres) lakes using the host-specific weevil *Cyrtobagous salviniae*. This eco-friendly intervention improved water quality, revived fisheries, and supported chestnut cultivation, directly enhancing local farmers' income. Her sustained efforts in promoting biocontrol technologies exemplify sustainable weed management and her commitment to farmers' social upliftment. She can be contacted through email id., anokhearchana12@gmail.com and archana.anokhe@icar.org.in

Dr. G. K. Dinesh (Govindaraj Kamalam Dinesh) serves as an Assistant Professor of Environmental Sciences at College of Agriculture, Central Agricultural University, Iroisemba, Imphal, India, under the Government of India. He earned his PhD in Environmental Science from ICAR-Indian Agricultural Research Institute, New Delhi, achieving a national-level 5th rank and receiving the IARI PhD Senior Research Fellowship through a national entrance examination. Additionally, he is a recipient of the 'Young Advantage Fellow' award from the Institute of Biosciences, USA, and holds an honorary Research Fellow position at INTI International University, Malaysia.

Dr. Dinesh completed his bachelor's degree in agriculture at Annamalai University, Chidambaram, Tamil Nadu, India, and his master's degree in (Agriculture) Environmental Science from Tamil Nadu Agricultural University, Coimbatore, Tamil Nadu, India. During his master's program, he received a fellowship for academic excellence through the Student Junior Research Fellowship under the TNAU-Private Industrial Agency Scheme (Sakthi Sugars Pvt. Ltd.) and conducted a long-term impact assessment study on distillery effluent application on soil and Cassava and Jasmine crops. His master's thesis focused on the ecology of birds and insects in organic and inorganic paddy ecosystems.

He has qualified for the ICAR and UGC National Eligibility Tests for Professorship multiple times. His PhD research on conservation agriculture and ecosystem services valuation was supported by the University Grants Commission, Government of India, through the UGC Junior Research Fellowship, and his work was recognized as significant by the Division of Environment Science, ICAR-Indian Agricultural Research Institute, India.

Dr. Dinesh's research expertise encompasses carbon sequestration, ecosystem services, agroecology, climate-smart agriculture, farming systems, sustainable agriculture, crop simulation modeling, GHG emissions, agrobiodiversity, bioremediation, economic ornithology, waste recycling, and pollution control. He has presented his research at numerous national and international conferences, earning accolades such as the 'Best Oral Paper Presentation Award,' 'Best MSc Thesis Award,' and 'Young Researcher Award for Excellence in Research.'

He has authored over 35 peer-reviewed publications, 15+ book chapters, and 70+ popular scientific articles in national and international journals and magazines. He has amassed over 300 citations, an h-index of 10, and a cumulative impact factor exceeding 35.

Beyond academia, Dr. Dinesh has translated numerous technical documents, including the Environmental Impact Assessment Draft by the Government of India. Since June 2024, he has served as a Subject Expert Member in the Book Selection Committee for Public Libraries under the Government of Tamil Nadu. He is also an expert member of various organizations, including the Food and Agricultural Organization, Rome, and has experience as a journalist and editorial member for a leading journals and magazine such as Enlighten Agriculture. His mission is to bridge the gap between science and society, fostering an inclusive and dialogic classroom environment through an initiative 'learn and grow together'. Currently, he is writing and editing books for prominent publishers such as CRC Press (UK), Nova Science (USA), and Bentham Science Publishers (UAE). He can be contacted at gkdineshiari@gmail.com or gkdineshiari@yahoo.com.

Index

#

α

β

λ

A

H

I

J

K

L

M

N

O

P